Lecture Notes in Computer Science 5

Commenced Publication in 1973
Founding and Former Series Editors:
Gerhard Goos, Juris Hartmanis, and Jan van

Alejandro López-Ortiz   Angèle Hamel (Eds.)

Combinatorial and Algorithmic Aspects of Networking

First Workshop on Combinatorial and
Algorithmic Aspects of Networking, CAAN 2004
Banff, Alberta, Canada, August 5-7, 2004
Revised Selected Papers

 Springer

Volume Editors

Alejandro López-Ortiz
University of Waterloo
School of Computer Science
200 University Ave. W.
Waterloo, Ontario N2L 3G1, Canada
E-mail: alopez-o@uwaterloo.ca

Angèle Hamel
Wilfrid Laurier University
Department of Physics and Computer Science
Waterloo, Ontario N2L 3C5, Canada
E-mail: ahamel@wlu.ca

Library of Congress Control Number: 2005929059

CR Subject Classification (1998): F.1.1, F.2.1-2, C.2, G.2.1-2, E.1

ISSN   0302-9743
ISBN-10   3-540-27873-7 Springer Berlin Heidelberg New York
ISBN-13   978-3-540-27873-3 Springer Berlin Heidelberg New York

Springer is a part of Springer Science+Business Media

springeronline.com

© Springer-Verlag Berlin Heidelberg 2005
Printed in Germany

Typesetting: Camera-ready by author, data conversion by Scientific Publishing Services, Chennai, India
Printed on acid-free paper   SPIN: 11527954   06/3142   5 4 3 2 1 0

Preface

The Internet is a massive global network of over 700 million users and it is adding users at the rate of 300,000 per day. This large, distributed, and everchanging network poses a challenge to researchers: How does one study, model, or understand such a decentralized, constantly evolving entity? Research in large-scale networks seeks to address this question, and the unique nature of these networks calls for a range of techniques from a host of disciplines. The workshop Combinatorial and Algorithmic Aspects of Networking and the Internet (CAAN 2004) provided a forum for the exchange of ideas on these topics.

The primary goals of the workshop were to bring together a diverse cross-section of researchers in an already scattered and distinct community and also to provide a snapshot of the cutting-edge research in this field. We succeeded in these goals: among the participants were mathematicians, computer scientists in theory and algorithms, computer scientists in networks, physicists, and engineers, as well as researchers from Europe and North America, participants from industry and academia, students, and established researchers; and among the papers were some new and surprising results as well as some introductions to the foundations of the field.

The workshop program featured 12 peer-reviewed papers bracketed by two hour-long invited survey talks—an opening talk by Ashish Goel and a closing talk by Andrei Broder. Topics covered by the talks ranged from the Web graph to game theory to string matching, all in the context of large-scale networks. This volume collects together the talks delivered at the workshop along with a number of survey articles to round out the presentation and give a comprehensive introduction to the topic.

We were fortunate to be given the opportunity to hold the conference as a two-day workshop at the Banff International Research Station for Mathematical Innovation and Discovery, BIRS. The breathtaking and inspiring setting and ample amenities contributed greatly to the success of the workshop. Attendance at BIRS is by invitation only and we had 24 participants at CAAN 2004. The small number of participants facilitated an intimate atmosphere perfect for generating discussions and initiating collaborations.

We would like to thank the Steering Committee for their guidance, and the Program Committee for their diligent work in reviewing the papers and selecting an excellent and balanced program. Special thanks goes to the organizational team at BIRS, especially Andrea Lundquist and Jackie Kler who made our job easy. We also thank Chris Taylor (http://photos.t-a-y-l-o-r.com) who graciously gave us permission to use the photographic artwork on the workshop poster. Finally, of course, we thank all the participants whose enthusiasm and support made CAAN 2004 a success and encouraged us to offer the workshop again.

CAAN 2005 will be held in August 2005 in Waterloo, Ontario, Canada as a satellite workshop of the Workshop on Algorithms and Data Structures (WADS 2005).

April 2005   Angèle Hamel and Alejandro López-Ortiz

Organization

Steering Committee

Andrei Broder	IBM, USA
Angèle Hamel	Wilfrid Laurier University, Canada
Srinivasan Keshav	University of Waterloo, Canada
Alejandro López-Ortiz	University of Waterloo, Canada
Rajeev Motwani	Stanford University, USA
Ian Munro	University of Waterloo, Canada

Program Committee

Azer Bestavros	Boston University, USA
Anthony Bonato	Wilfrid Laurier University, Canada
Andrei Broder	IBM Research, USA
Hervé Brönnimann	Polytechnic University, USA
Adam Buchsbaum	AT&T Labs, USA
Edith Cohen	AT&T Labs, USA
Erik Demaine	MIT, USA
Luisa Gargano	University of Salerno, Italy
Ashish Goel	Stanford University, USA
Angèle Hamel (Org. Comm. Chair)	Wilfrid Laurier University, Canada
Monika Henzinger	Google, USA
Jeannette Janssen	Dalhousie University, Canada
David Karger	MIT, USA
Srinivasan Keshav	University of Waterloo, Canada
Alejandro López-Ortiz (Program Chair)	University of Waterloo, Canada
Bruce Maggs	Carnegie Mellon University, USA
Rajeev Motwani	Stanford University, USA
Ian Munro	University of Waterloo, Canada
Tim Roughgarden	University of California at Berkeley, USA
Christos Papadimitrou	University of California at Berkeley, USA
David Peleg	Weizmann Institute, Israel
Torsten Suel	Polytechnic University, USA
Eli Upfal	Brown University, USA
Alessandro Vespignani	Laboratoire de Physique Théorique, France

Table of Contents

Workshop Proceedings

Survey Articles

Aggregating Correlated Data in Sensor Networks

M. Enachescu[1], A. Goel[1], R. Govindan[2], and R. Motwani[1]

[1] Stanford University, California, USA
[2] University of Southern California, California, USA

Abstract for Invited Presentation. Consider a network where each node gathers information from its vicinity and sends this information to a centralized processing agent. If the information is geographically correlated, then a large saving in data transmission costs may be obtained by aggregating information from nearby nodes before sending it to the central agent. This is particularly relevant to sensor networks where battery limitations dictate that data transmission be kept to a minimum, and where sensed data is often geographically correlated. In-network aggregation for sensor networks has been extensively studied over the last few years. In this paper we show that a very simple opportunistic aggregation scheme can result in near-optimum performance under widely varying (and unknown) scales of correlation.

More formally, we consider the idealized setting where sensors are arranged on an $N \times N$ grid, and the centralized processing agent is located at position $(0,0)$ on the grid. We assume that each sensor can communicate only to its four neighbors on the grid. This idealized setting has been widely used to study broad information processing issues in sensor networks. We call an aggregation scheme *opportunistic* if data from a sensor to the central agent is always sent over a shortest path, i.e., no extra routing penalty is incurred to achieve aggregation.

To model geographic correlations, we assume that each sensor can gather information in a $\frac{k}{2} \times \frac{k}{2}$ square (or, a circle of radius $k/2$) centered at the sensor. We will refer to k as the correlation parameter. Let the set $A(x)$ denote the area sensed by sensor i. If we aggregate information from a set of sensors S then the size of the resulting compressed information is $I(S) = \left| \bigcup_{x \in S} A(x) \right|$, i.e., the size of the total area covered by the sensors in S. Often, the parameter k can depend on the intensity of the information being sensed. For example, a volcanic eruption might be recorded by many more sensors and would correspond to a much higher k than a campfire. Accordingly, we will assume that the parameter k is not known in advance. In fact, we would like our opportunistic aggregation algorithms to work well simultaneously for all k.

There are scenarios where the above model applies directly. For example, the sensors could be cameras which take pictures within a certain radius, or they could be sensing RFID tags on retail items (or on birds which have been tagged for environmental monitoring) within a certain radius. Also, since we want algorithms that work well without any knowledge of k, our model applies to scenarios where the likelihood of sensing decreases with distance. Thus, we believe that our model

A. López-Ortiz and A. Hamel (Eds.): CAAN 2004, LNCS 3405, pp. 1–2, 2005.

(optimizing simultaneously for all k) captures the joint entropy of correlated sets of sensors in a natural way for a large variety of applications.

For node (i, j), we will refer to nodes $(i-1, j)$ and $(i, j-1)$ as its downstream neighbors, and nodes $(i+1, j)$ and $(i, j+1)$ as its upstream neighbors. We would like to construct a tree over which information flows to the central agent, and gets aggregated along the way. Since we are restricted to routing over shortest paths, each node has just one choice: which downstream node to choose as its parent in the tree. Each node aggregates the information it sensed locally with any information it received from its upstream neighbors and sends it on to one of its downstream neighbors. The cost of the tree is the total amount of (compressed) information sent out over links in the tree.

Our Results: We propose a very simple randomized algorithm for choosing the next neighbor – node (i, j) chooses its left neighbor with probability $i/(i+j)$ and its bottom neighbor with probability $j/(i+j)$. Observe that this scheme results in all shortest paths between (i, j) and $(0, 0)$ being chosen with equal probability. We prove that this simple scheme is a constant factor approximation (in expectation) to the optimum aggregation tree *simultaneously* for all correlation parameters k. While we construct a single tree, the optimum trees for different correlation parameters may be different. The key idea is to relate the expected collision time of random walks on the grid to scale free aggregation.

Our results hold only for the total cost, and critically rely on the fact that information is distributed evenly through the sensor field. This result shows that, at least for the class of aggregation functions and the grid topology considered in this paper, schemes that attempt to construct specialized routing structures in order to improve the likelihood of data aggregation are unnecessary. This is convenient, since such specialized routing structures are hard to build without some a priori knowledge about correlations in the data. With this result, simple geographic routing schemes or tree-based data gathering protocols are sufficient.

The Efficiency of Optimal Taxes

George Karakostas[1],* and Stavros G. Kolliopoulos[2],**

[1] Department of Computing and Software, McMaster University
karakos@mcmaster.ca
[2] Department of Informatics and Telecommunications,
University of Athens and Department of Computing and Software,
McMaster University
www.cas.mcmaster.ca/~stavros

Abstract. It is well known that the selfish behavior of users in a network can be regulated through the imposition of the so-called *optimal taxes* on the network edges. Any traffic equilibrium reached by the selfish users who are conscious of both the travel latencies and the taxes will minimize the social cost, i.e., will minimize the total latency.

Optimal taxes incur desirable behavior from the society point of view but they cause disutility to the network users since the users' total cost is in general increased [4]. Excessive disutility due to taxation may be undesirable from the societal perspective as well. In this work we examine the efficiency of taxation as a mechanism for achieving the desired goal of minimizing the social cost. We show that for large classes of latency functions the total disutility due to taxation that is caused to the users and/or the system is bounded with respect to the social optimum. In addition, we show that if the social cost takes into account both the total latency and the total taxation in the network, the coordination ratio for certain latency functions is better than the coordination ratio when taxation is not used.

1   Introduction

In the *selfish routing* setting, we are given a directed network $G = (V, E)$ and a set of k *classes of users* (commodities), each with its own origin and destination, and with a fixed total demand (traffic) rate per class $d_i > 0$, $i = 1, \ldots, k$. Individual users are thought as carrying each an infinitesimal amount of a commodity. We are also given a nonnegative latency function l_P describing the delay experienced by users wishing to travel on the path P as a function of the total flow through the edges of the path. In this work we assume that the *additive model* holds, i.e., for every edge e there is a latency function $l_e(f_e)$ that describes the latency on this edge due to the flow f_e that crosses it. Then the latency for a path is defined as $l_P(f) := \sum_{e \in P} l_e(f_e)$. Each user tries to selfishly route his flow so

* Research supported by an NSERC Discovery grant.
** Research partially supported by NSERC Research Grant 227809-00.

A. López-Ortiz and A. Hamel (Eds.): CAAN 2004, LNCS 3405, pp. 3–12, 2005.

that his path cost is minimized. A *traffic equilibrium* is an assignment of traffic to paths so that no user can unilaterally switch her flow to a path of smaller cost. Wardrop's principle [16] for selfish routing postulates that

at equilibrium, for each origin-destination pair, the travel costs on all the routes actually used are equal, or less than the travel costs on all nonused routes.

A widely used measure of the network performance is the *social cost* (or *total latency*), defined as $\sum_{\text{path } P} f_P l_P(f) = \sum_{e \in E} f_e l_e(f_e)$ for a flow f that routes f_P units of traffic through path P. Although it must obey Wardrop's principle at equilibrium, the unregulated choice of paths by individual users may incur a social cost which in general can be higher than the social optimum. In fact the latency of an equilibrium can be arbitrarily larger than the social optimum [14]. A classic way of dealing with this problem is the introduction of *taxation* on the edges of the network, so that the users' path cost has both a travel time and a budgetary component. Without taxation, users experience only their own traffic delay as their cost. With taxation users are also charged for the right to use a path. This technique has been studied by the traffic community for a long time (cf. [5] and the references therein), especially in the context of *marginal costs* (see, for example, [2],[8],[15]). Each selfish user of class i using path P will experience the following path cost:

$$\text{path cost}(P) := \text{latency}(P) + a(i) \cdot \text{taxation}(P).$$

The taxation(P) is the sum of taxes along the edges of the path. The factor $a(i) > 0$, denotes the sensitivity of user class i to the taxes. In the *homogeneous* case all user classes have the same sensitivity to the taxation (i.e. $a(i) = 1$, for all i), while in the *heterogeneous* case $a(i)$ can take different positive values for different classes. Through edge taxation, we would like to force *all* equilibria on the network to induce flow that minimizes the social cost $\sum_{e \in E} f_e l_e(f_e)$. We refer to a set of edge taxes that achieves this as *optimal taxes*. In the homogeneous case, marginal costs have been shown (cf. [2],[8],[15]) to be optimal taxes. In the heterogeneous case, the existence and calculation of such optimal taxes were shown for the single source-destination pair case by Cole et a. [5], and were later extended to the multicommodity setting in [9, 11].

Designing optimal taxes is a classic instance of mechanism design, a central topic in game theory (see, e.g., [13]). A set of outcomes is fixed (here achieving the social optimum) and users are paid or penalized, (here they pay taxes) in order to achieve the desired outcome in equilibrium. One can actually see the taxation cost in two different contexts. One is the context already discussed, which is as monetary cost. The other is to see the tax for every edge as part of the edge latency function itself. Then, instead of taxation, we can speak about *artificial delays* introduced possibly at the entrance of each edge, in order to minimize the total amount of time users actually spend on the edges themselves. For example, this is the technique used at some highway exits, where traffic lights have been installed in order to better control traffic. Whatever the context though, taxation increases in general the user cost, as was shown in [4] for the case of marginal cost taxes. The natural question that arises is whether taxes are an efficient mechanism for achieving the desired result. Is the additional disutility caused through taxation proportionate to the desired goal, i.e., a routing that minimizes the total latency? In this paper we tackle this problem by comparing the social cost of the traffic equilibria when taxation is used against (i) the social optimum without taxation and (ii) the social optimum when taxation is taken into account.

In Section 4 we show that in the homogeneous case the ratio of social cost at equilibrium with taxation to the social optimum without taxation is not much bigger than the worst case ratio without any taxation for many important families of latency functions, like linear or low-degree polynomial ones. In particular for strictly increasing linear latency functions we show that, if b is the vector of optimal edge taxes (in this case, the marginal cost taxes),

$$\frac{\sum_e f_e^*(l_e(f_e^*) + b_e)}{\sum_e \hat{f}_e l_e(\hat{f}_e)} \leq 2$$

for any equilibrium flow f^* and flow $\hat{f}$ that achieves the social optimum. This bound is tight, and is not far from the $4/3$ upper bound on the *coordination ratio* in the case without taxes, shown by Roughgarden and Tardos [14]. The coordination ratio ρ was defined by Koutsoupias and Papadimitriou in [12] as follows

$$\rho := \sup_{f^*} \frac{\sum_e f_e^* l_e(f_e^*)}{\sum_e \hat{f}_e l_e(\hat{f}_e)}$$

for the worst case (in terms of social cost) equilibrium f^*, and $\hat{f}$ as before. Hence we show that with a small increase in network inefficiency, we achieve, at equilibrium, a flow pattern that minimizes the total latency of the users. Note that, in principle, the tax b_e on an edge e could be very big compared to the latency part $l_e()$ of the edge cost function. Hence it is rather surprising that taxation does not drive the social cost further than a small constant factor away from the social (without taxation) optimum.

This approach in bounding the inefficiency of taxation as a mechanism to achieve minimum social cost is influenced by the notion of coordination mechanisms. This concept was recently introduced by Christodoulou, Koutsoupias and Nanavati [3]. Informally speaking a coordination mechanism is a cost function experienced by the users, chosen from a family of possible cost functions called a coordination model. The measure of the efficiency of a coordination mechanism is the supremum over all possible demand sets of the ratio of the social cost of the worst-case equilibrium to the social optimum achieved with the *original* cost function. See [3] for the precise mathematical definitions. Note that in our work the demands are fixed.

Proving that taxation incurs a small increase to the cost of an equilibrium compared to the social optimum without taxation is a satisfying result. However it is possible that once taxes are imposed by some central authority they are considered to be part of the social cost. In other words taxation may incur disutility to society as a whole. To address this issue, we compare the worst-case cost with taxation of an equilibrium against the social optimum with taxation. In other words we consider the coordination ration in the standard sense [12] of the game with taxes. In Section 3 we show that, for certain families of strictly increasing and continuous latency functions (like linear or polynomial ones), the coordination ratio of the network actually *decreases* when optimal taxes are introduced. In particular for the linear latency functions case, we show that

$$\frac{\sum_e f_e^*(l_e(f_e^*) + b_e)}{\sum_e \bar{f}_e(l_e(\bar{f}_e) + b_e)} \leq \frac{5}{4}$$

for any equilibrium flow f^* and social optimum $\bar{f}$ when taxes b are used in both cases. This is significantly better than the $4/3$ bound of [14] for general linear functions. The gap between $4/3$ and $5/4$ quantifies the beneficial effect of taxation on the behavior of the selfish users, specifically the reduction in their resistance to coordination. Our bound holds for heterogeneous users as well, and its proof is based on the definition of the parameter $\beta(\mathcal{L})$ for a family of functions $\mathcal{L}$ by Correa, Schulz and Stier Moses [6].

The two results in combination show that imposing the optimal taxes on a selfish routing game not only yields a game with better coordination ratio, but also the added disutility to the users is bounded with respect to the original system optimum. In addition we emphasize that our two approaches together provide a stronger guarantee on the worst-case cost of an equilibrium with taxation than each one of them taken separately. For a given tax vector b, let $\bar{f}$ be the social optimum with taxation and $\hat{f}$ be the social optimum without taxation. There does not seem to be any a priori information about which of the two quantities

$$\frac{5}{4}\sum_e \bar{f}_e(l_e(\bar{f}_e) + b_e), \ \ 2\sum_e \hat{f}_e l_e(\hat{f}_e)$$

is smaller.

2   The Model

Let $G = (V, E)$ be a directed network (possibly with parallel edges but with no self-loops), and a set of *users*, each with an infinitesimal amount of traffic (flow) to be routed from an origin node to a destination node of G. Moreover, each user α has a positive *tax-sensitivity* factor $a(\alpha) > 0$. We will assume that the tax-sensitivity factors for all users come from a finite set of possible positive values. We can bunch together into a single *user class* all the users with the same origin-destination pair and with the same tax-sensitivity factor; let k be the number of different such classes. We denote by $d_i, \mathcal{P}_i, a(i)$ the total flow of class i, the flow paths that can be used by class i, and the tax-sensitivity of class i, for all $i = 1, \ldots, k$ respectively. We will also use the term 'commodity i' for class i. Set $\mathcal{P} \doteq \cup_{i=1,\ldots,k}\mathcal{P}_i$. Each edge $e \in E$ is assigned a *latency function* $l_e(f_e)$ which gives the latency experienced by any user that uses e due to congestion caused by the total flow f_e that passes through e. In other words, as in [5], we assume the additive model in which for any path $P \in \mathcal{P}$ the latency is $l_P(f) = \sum_{e \in P} l_e(f_e)$, where $f_e = \sum_{e \ni P} f_P$ and f_P is the flow through path P. If every edge is assigned a per-unit-of-flow tax $\beta_e \geq 0$, a selfish user in class i that uses a path $P \in \mathcal{P}_i$ experiences total cost of

$$\sum_{e \in P} l_e(f_e) + a(i) \sum_{e \in P} \beta_e$$

hence the name 'tax-sensitivity' for the $a(i)$'s: they quantify the importance each user assigns to the taxation of a path.

Let $\hat{f}$ be a flow that satisfies the users' demands and minimizes the social cost $\sum_{e \in E} f_e l_e(f_e) = \sum_i \sum_{P \in \mathcal{P}_i} f_P l_P(f)$, i.e., $\hat{f}$ is a solution of the following mathematical program:

$$\min \sum_{e \in E} f_e l_e(f_e) \quad \text{s.t.} \qquad \text{(MP)}$$

$$\sum_{P \in \mathcal{P}_i} f_P = d_i \qquad \forall i$$

$$f_e = \sum_{P \in \mathcal{P}: e \in P} f_P \quad \forall e \in E$$

$$f_P \geq 0 \qquad \forall P$$

Note that, although in certain cases (e.g. when the latency functions l_e are convex) the flow $\hat{f}$ can be computed efficiently, for more general latency functions it may be extremely difficult to compute $\hat{f}$ (see Section 4 in [5]). We will assume that such an $\hat{f}$ exists and that it induces finite latency on every edge.

A function $g(x)$ is *positive* if $g(x) > 0$ when $x > 0$. We assume that the functions l_e are strictly increasing, i.e., $x > y \geq 0$ implies $l_e(x) > l_e(y)$, and that $l_e(0) \geq 0$. This implies that $l_e(f_e) > 0$ when $f_e > 0$, i.e., the function l_e is positive. Similar assumptions on monotonicity are made in [5].

Let

$$K := \{f : 0 \leq f_P, \forall P \wedge \sum_{P \in \mathcal{P}_i} f_P = d_i, \forall i\}$$

be the set of all flows that satisfy the users' demands.

Definition 1. *A flow f is called* feasible *iff $f \in K$.*

A traffic (or Wardrop) equilibrium is a feasible flow $f^* \in K$ such that

$$\langle T(f^*), f - f^* \rangle \geq 0, \quad \forall f \in K. \qquad (1)$$

where $\langle \cdot, \cdot \rangle$ denotes the inner product, and $T_P(f)$ is the function that gives the generalized cost of a user that uses path P when the network flow is f.

3   Improving the Coordination Ratio

By extending the results of [5] from the single source-sink to the multicommodity setting in [11], we have shown that there is a set of per-unit taxes b which forces the users to induce on the edges the same edge flow as $\hat{f}$. This result was also independently obtained by Fleischer, Jain and Mahdian [9].

Theorem 1. (Theorem 1 in [11]) *Consider the selfish routing game with the latency function seen by the users in class i being*

$$T_P(f) := \sum_{e \in E} l_e(f_e) + a(i) \sum_{e \in P} b_e, \quad \forall i, \; \forall P \in \mathcal{P}_i.$$

If for every edge $e \in E$ l_e is a strictly increasing function with $l_e(0) \geq 0$, then there is a vector of per-unit taxes $b \in \mathbb{R}_+^{|E|}$ such that, if f^* is a traffic equilibrium for this game, $f_e^* = \hat{f}_e$, $\forall e \in E,$. Therefore f^* minimizes the social cost $\sum_{e \in E} f_e l_e(f_e)$.

We also can compute this b in polynomial time when $\hat{f}$ is given to us.

Correa et al. [6, 7] define the following quantity β for a continuous nondecreasing latency function $l : \mathbb{R}^+ \to \mathbb{R}^+$ and every value $u \geq 0$:

$$\beta(u, l) := \frac{1}{u l(u)} \max_{x \geq 0}\{x(l(u) - l(x))\}, \tag{2}$$

where by convention $0/0 = 0$. In addition, they define $\beta(u) := \sup_{u \geq 0} \beta(u, l)$ and $\beta(\mathcal{L}) := \sup_{l \in \mathcal{L}} \beta(l)$, where $\mathcal{L}$ is a family of latency functions. Note that $\beta(l) \geq \beta(u, l)$, $\forall u, l$, and $\beta(\mathcal{L}) \geq \beta(l)$, $\forall l \in \mathcal{L}$. Also $0 \leq \beta(\mathcal{L}) < 1$.

Define the *cost* $C(f)$ of flow f to be $C(f) := \sum_P f_P T_P(f)$, and assume the *additive model* for l, i.e., $l_P(f) := \sum_{e \in P} l_e(f_e)$, where l_e is a function that gives the latency for edge e when flow f_e passes through it.

Theorem 2. *Let f^* be any traffic equilibrium for the game of Theorem 1, and $\bar{f}$ be a flow that minimizes $C(f)$. Then, for latency functions l_e that belong to a family $\mathcal{L}$*

$$\frac{C(f^*)}{C(\bar{f})} \leq 1 + \beta(\mathcal{L}).$$

Proof. First we note that $C(f^*) = \langle T(f^*), f^* \rangle$. Also, because of Theorem 1 we have that $f_e^* = \hat{f}_e$. Then, for any feasible flow $\bar{f} \in K$ we have

$$\langle T(f^*), \bar{f} \rangle = \sum_i \sum_{P \in \mathcal{P}_i} \bar{f}_P \left(l_P(f^*) + a(i) \sum_{e \in P} b_e \right)$$

$$= \sum_{e \in E} \bar{f}_e l_e(f_e^*) + \sum_i \sum_{P \in \mathcal{P}_i} \bar{f}_P a(i) \sum_{e \in P} b_e \tag{3}$$

$$= \sum_{e \in E} \bar{f}_e l_e(\hat{f}_e) + \sum_i \sum_{P \in \mathcal{P}_i} \bar{f}_P a(i) \sum_{e \in P} b_e$$

where the second equality is due to the additive model.

We also have by definition that

$$\beta(\hat{f}_e, l_e) := \frac{1}{\hat{f}_e l_e(\hat{f}_e)} \max_{x \geq 0} \{x(l_e(\hat{f}_e) - l_e(x))\},$$

from which we get for $x := \bar{f}_e$ that

$$\beta(\hat{f}_e, l_e)\hat{f}_e l_e(\hat{f}_e) \geq \bar{f}_e l_e(\hat{f}_e) - \bar{f}_e l_e(\bar{f}_e)$$

or, by the definition of $\beta(\mathcal{L})$,

$$\beta(\mathcal{L})\hat{f}_e l_e(\hat{f}_e) \geq \bar{f}_e l_e(\hat{f}_e) - \bar{f}_e l_e(\bar{f}_e). \tag{4}$$

By the definition of $\hat{f}$ we know that $\sum_{e \in E} \hat{f}_e l_e(\hat{f}_e) \leq \sum_{e \in E} \bar{f}_e l_e(\bar{f}_e)$, therefore (4) implies that

$$(1 + \beta(\mathcal{L})) \cdot \sum_{e \in E} \bar{f}_e l_e(\bar{f}_e) \geq \sum_{e \in E} \bar{f}_e l_e(\hat{f}_e),$$

and this, in turn, together with (3) implies that

$$\langle T(f^*), \bar{f} \rangle \leq (1 + \beta(\mathcal{L})) \cdot \sum_{e \in E} \bar{f}_e l_e(\bar{f}_e) + \sum_{i} \sum_{P \in \mathcal{P}_i} \bar{f}_P a(i) \sum_{e \in P} b_e$$

$$\leq (1 + \beta(\mathcal{L})) \cdot \sum_{i} \sum_{P \in \mathcal{P}_i} \bar{f}_P \left(l_P(\bar{f}) + a(i) \sum_{e \in P} b_e \right) \tag{5}$$

$$= (1 + \beta(\mathcal{L})) \cdot C(\bar{f}).$$

Since f^* is an equilibrium, (1) implies that

$$\langle T(f^*), f \rangle \geq \langle T(f^*), f^* \rangle, \quad \forall f \in K,$$

which, together with (5) and for $f := \bar{f}$, implies that

$$(1 + \beta(\mathcal{L})) \cdot C(\bar{f}) \geq C(f^*).$$

Correa et al. [6] give upper bounds of $\beta(\mathcal{L})$ for some function families $\mathcal{L}$. We repeat here these bounds (Corollaries 4.3, 4.4 in [6]):

Corollary 1. [6] *If the set $\mathcal{L}$ of continuous and nondecreasing latency functions is contained in the set $\{l(\cdot) : l(cx) \geq cl(x) \text{ for } c \in [0,1]\}$, then $\beta(\mathcal{L}) \leq 1/4$.*

Note that Corollary 1 holds for the family of nondecreasing linear functions, hence the worst case bound of Theorem 2 for linear functions is 5/4 which is better than the tight 4/3 bound [14] achieved without the use of the b's from Theorem 1.

Corollary 2. [6] *If the set $\mathcal{L}$ of continuous and nondecreasing latency functions is contained in the set $\{l(\cdot) : l(cx) \geq c^n l(x) \text{ for } c \in [0,1]\}$ for some positive number n, then*

$$\beta(\mathcal{L}) \leq \frac{n}{(n+1)^{1+1/n}}.$$

Therefore for $n = 2, 3$ and 4 the upper bound becomes $1.385, 1.472$ and 1.535 respectively, as opposed to $1.626, 1.896$ and 2.151 respectively when b is not used [6]. Also note that as n increases, the bound goes to 2.

4   Comparison to the Original Latencies

In general, the values of b in Theorem 1 can be very big. It may even be the case that the part of the cost $C(f)$ due to the initial latencies $\sum_e f_e l_e(f_e)$ is negligible compared to the part due to b, which is $\sum_i \sum_P a(i) f_P \sum_{e \in P} b_e$. Therefore the improvement of the coordination ratio may come at a prohibitive increase to the overall cost. One would like to bound b so that the new overall cost is bounded by a function of the original optimal total latency $\sum_e \hat{f}_e l_e(\hat{f}_e)$.

Unfortunately, we do not know how to bound b for the general $l(\cdot)$ of Theorem 1. But we can use already known results in the case of *homogeneous* users, i.e., $a(i) = 1$, $\forall i$, to bound the ratio of the worst equilibrium cost when b is used to the original optimal total cost.

It is well known ([2],[8],[15]; see also [4], especially Proposition 3.1) that, for homogeneous networks with differentiable latency functions l_e, one can use the *marginal costs*[1] $\hat{f}_e l_e'(\hat{f}_e)$ as b_e in Theorem 1 to achieve the following classical result:

Theorem 3. *If functions l_e are differentiable, then $\hat{f}$ is an equilibrium for the selfish routing game with*

$$T_P(f) := \sum_{e \in P} (l_e(f_e) + \hat{f}_e l_e'(\hat{f}_e)).$$

Moreover, if we assume that l_e are strictly increasing (as in Theorem 1), then any equilibrium f^* incurs the same edge flow as $\hat{f}$, i.e., $f_e^* = \hat{f}_e$, $\forall e$ (Theorem 6.2 in [1]). Let $C_{OPT} := \sum_{e \in E} \hat{f}_e l_e(\hat{f}_e)$.

Theorem 4. *If $l_e(f_e) = a_e f_e + \beta_e$ with $a_e > 0, b_e \geq 0$ for all $e \in E$, and f^* is an equilibrium for the selfish routing game with taxes $b_e := a_e \hat{f}_e$, then*

$$\frac{C(f^*)}{C_{OPT}} \leq 2.$$

More generally, if the l_e's are polynomials of degree d with positive coefficients, then

$$\frac{C(f^*)}{C_{OPT}} \leq d + 1.$$

Proof. For the case of linear latency functions, note that these functions are differentiable, therefore Theorem 3 implies that $\hat{f}$ is an equilibrium. Since they are strictly increasing as well, we know that $f_e^* = \hat{f}_e$, $\forall e$, for any equilibrium f^*. Hence

$$C(f^*) := \sum_P f_P^* T_P(f^*) = \sum_e f_e^* (a_e f_e^* + \beta_e + a_e \hat{f}_e)$$

$$= \sum_e \hat{f}_e (2 a_e \hat{f}_e + \beta_e)$$

$$\leq 2 \cdot C_{OPT}$$

The same argument shows the upper bound for the case of degree d polynomials.

---

[1] $l_e'(\cdot)$ is the derivative of $l_e(\cdot)$.

Note that the bound above is tight in the case of polynomials with all the coefficients, except for the one of the highest degree factor, being 0.

Theorem 5. *If the l_e functions are strictly increasing and continuously differentiable with a convex derivative, and f^* is an equilibrium for the selfish routing game with taxes $b_e := a_e \hat{f}_e$, then*

1. *If $l_e(\cdot) \in \{l(\cdot) : l(cx) \geq cl(x) \text{ for } c \in [0,1]\}$, then*

$$\frac{C(f^*)}{C_{OPT}} \leq 3,$$

2. *If $l_e(\cdot) \in \{l(\cdot) : l(cx) \geq c^n l(x) \text{ for } c \in [0,1]\}$, then*

$$\frac{C(f^*)}{C_{OPT}} \leq 2^n + 1.$$

Proof. Theorem 3 implies that $\hat{f}$ is an equilibrium. Since the l_e's are strictly increasing as well, we know that $f_e^* = \hat{f}_e$, $\forall e$, for any equilibrium f^*. Observe that the edges e for which $\hat{f}_e = 0$ do not contribute to $C(f^*)$ therefore we ingore them in the ensuing calculations. It is known that if the l_e' functions are convex and continuous in (γ, δ), then

$$l_e'(x) \leq \frac{1}{2h} \int_{x-h}^{x+h} l_e'(t)dt \tag{6}$$

for $\gamma \leq x - h < x < x + h \leq \delta$ (fact 125, p. 98 of [10]). For $x := f_e/2, h := f_e/2$, inequality (6) becomes

$$f_e l_e'(\frac{f_e}{2}) \leq \int_0^{f_e} l_e'(t)dt. \tag{7}$$

Under the assumptions of Part 1 of the theorem, (7) gives

$$f_e l_e'(f_e) \leq 2l_e(f_e)$$

which, for $f_e := \hat{f}_e$, implies together with an argument similar to the proof of Theorem 4 that

$$\frac{C(f^*)}{\sum_e \hat{f}_e l_e(\hat{f}_e)} \leq 3$$

for any equilibrium flow f^*.

Under the assumptions of Part 2 of the theorem, (7) implies that

$$f_e l_e'(f_e) \leq 2^n l_e(f_e)$$

which, in turn, implies that in this case

$$\frac{C(f^*)}{\sum_e \hat{f}_e l_e(\hat{f}_e)} \leq 2^n + 1$$

for any equilibrium flow f^*.

Acknowledgement

Thanks to Elias Koutsoupias for useful discussions.

References

1. H. Z. Aashtiani and T. L. Magnanti. Equilibria on a congested transportation network. *SIAM Journal of Algebraic and Discrete Methods*, 2:213–226, 1981.
2. M. Beckmann, C. B. McGuire, and C. B. Winsten. *Studies in the Economics of Transportation*. Yale University Press, 1956.
3. G. Christodoulou, E. Koutsoupias, and A. Nanavati. Coordination mechanisms. To appear in Proc. 31st ICALP, 2004.
4. R. Cole, Y. Dodis, and T. Roughgarden. How much can taxes help selfish routing? In *Proceedings of the 4th ACM Conference on Electronic Commerce*, pages 98–107, 2003.
5. R. Cole, Y. Dodis, and T. Roughgarden. Pricing network edges for heterogeneous selfish users. In *Proceedings of the 35th Annual ACM Symposium on Theory of Computing*, pages 521–530, 2003.
6. J. R. Correa, A. S. Schulz, and N. E. Stier Moses. Selfish routing in capacitated networks. Technical Report Working Paper 4319-03, MIT Sloan School of Management, Cambridge, MA, June 2003.
7. J. R. Correa, A. S. Schulz, and N. E. Stier Moses. Selfish routing in capacitated networks. To appear in Mathematics of Operations Research, February 2004.
8. S. Dafermos and F. T. Sparrow. The traffic assignment problem for a general network. *Journal of Research of the National Bureau of Standards, Series B*, 73B:91–118, 1969.
9. L. Fleischer, K. Jain, and M. Mahdian. Taxes for heterogeneous selfish users in a multicommodity network. To appear in *Proceedings of the 45th Annual IEEE Symposium on Foundations of Computer Science*, 2004.
10. G. Hardy, J. E. Littlewood, and G. Pólya. *Inequalities - Second Edition*. Cambridge, 1934.
11. G. Karakostas and S. G. Kolliopoulos. Edge pricing of multicommodity networks for heteregoneous selfish users. To appear in *Proceedings of the 45th Annual IEEE Symposium on Foundations of Computer Science*, 2004.
12. E. Koutsoupias and C. Papadimitriou. Worst-case equilibria. In *Proceedings of the 16th Annual Symposium on Theoretical Aspects of Computer Science*, pages 404–413, 1999.
13. M. J. Osborne and A. Rubinstein. *A course in Game Theory*. The MIT Press, Cambridge, MA, 1994.
14. T. Roughgarden and É. Tardos. How bad is selfish routing? *Journal of the ACM*, 49:236–259, 2002.
15. M. J. Smith. The marginal cost taxation of a transportation network. *Transportation Research*, 13B:237–242, 1979.
16. J. G. Wardrop. Some theoretical aspects of road traffic research. *Proc. Inst. Civil Engineers, Part II*, 1:325–378, 1952.

Congestion Games, Load Balancing, and Price of Anarchy*

Anshul Kothari[1], Subhash Suri[1], Csaba D. Tóth[1], and Yunhong Zhou[2]

[1] Computer Science Depart., University of California,
Santa Barbara, CA 93106
{kothari, suri, toth}@cs.ucsb.edu
[2] HP Labs, 1501 Page Mill Rd, Palo Alto, CA 94304
yunhong.zhou@hp.com

Abstract. Imagine a set of self-interested clients, each of whom must choose a server from a permissible set. A server's latency is inversely proportional to its speed, but it grows linearly with (or, more generally, as the pth power of) the number of clients matched to it. Many emerging Internet-centric applications such as peer-to-peer networks, multi-player online games and distributed computing platforms exhibit such interaction of self-interested users. This interaction is naturally modeled as a congestion game, which we call *server matching*. In this overview paper, we summarize results of our ongoing work on the analysis of the server matching game, and suggest some promising directions for future research.

1   Introduction

The decentralized nature of the Internet has added a new dimension of complexity to the design and analysis of algorithms: *lack of cooperation or coordination*. The users of the Internet are a diverse group, each selfishly and independently interested in optimizing its own personal utility. These users are neither *obedient*, as typically assumed in theory of algorithms, nor *malicious*, as assumed in areas like cryptography. Instead, these selfish users are best modeled as *strategic* and *rational* entities, whose primary objective is to maximize their own benefit. For instance, they do not intentionally act to inflict harm on a system, but they also cannot be expected to follow a protocol truthfully if deviating from the protocol is to their advantage.

There has been much interest lately in understanding the implications of such selfish behavior in distributed systems that lack a central authority. Examples include the work on selfish routing [22, 23], traffic allocation [5], network formation [1, 10] etc. A good starting point in this quest is the following question: in a system of self-interested agents, each with its own private objective function,

---

* Research by the first three authors was partially supported by National Science Foundation grants CCR-0049093 and IIS-0121562.

A. López-Ortiz and A. Hamel (Eds.): CAAN 2004, LNCS 3405, pp. 13–27, 2005.

what is an optimal solution? A solution that optimizes the system's objective as a whole (also called the *social welfare*) is often unstable and thus untenable because some agents may improve their *own* payoff by changing their actions (strategies). Thus, the only reasonable solutions are those that are *stable*: solutions where no single agent has an incentive to deviate, the so-called *Nash equilibria*.

Not surprisingly, a Nash equilibrium solution, in general, can be arbitrarily worse than the (coordinated) optimal. In the traditional style of theoretical computer science, we use the ratio between the *worst-case* Nash solution to the optimal to denote the quality of a stable solution. Papadimitriou [19] has coined the phrase "price of anarchy" to denote this ratio. In this paper, we consider a natural problem of assigning (selfish) clients to servers in the absence of central coordination. Our problem setting is motivated by many Internet-centric applications where users independently choose servers, and the interaction of these self-interested users is naturally modeled by the game we consider. Examples of such interaction include users downloading music files in peer-to-peer networks [13, 12, 15], server selection in massively multi-player online games [9], distributed computing platforms [4], distributed testbeds [20], or users downloading software from repositories etc. Our problem is related to a classical (coordinated) version of load balancing problem [2], however, the lack of coordination and selfish clients add a new twist to that problem. We begin with a formal definition of our server matching game.

1.1   Our Model

Consider a set U of n clients and a set V of m servers. There is a (incomplete) bipartite graph G between U and V and a server v_j is *permissible* for client u_i only if (u_i, v_j) is an edge in G. Every client selfishly chooses a server in the absence of a coordinating authority. Each of them wants to minimize its latency (job completion time), and rationally prefers a fast server to a slower one. Each server's latency, however, is an increasing function of the server load (the number of clients served by it), and so depends on how many other clients also choose that same server. That is, each server v has a *load dependent*[1] monotone increasing latency function $\lambda_v : \mathbb{N} \to \mathbb{R}^+$.

A *server matching* is a mapping $M : U \to V$ that assigns each client to a server.[2] The number of clients assigned to server v in a (many to one) matching M is denoted by $d_M(v)$, the *load* of v in M. When the matching M is clear

---

[1] In general, a client's latency has two components: one server dependent and one network dependent. In our simplified model, we consider just the server latency and treat network delay to be constant. Our server latency model reflects the practical situation of a server multiplexing data transmission to its clients, and so the latency to each client grows with the number of clients.

[2] That is, each client is matched to exactly one server, but a server can be matched to multiple clients (or none). To avoid trivialities, we assume that for each client u there is at least one server that can provide the data, and each server has data for at least one client.

from the context, we will simply use the shortened notation $d(v)$. Each client matched to v in the matching M experiences a latency $\lambda_v(d_M(v))$. The *cost* of a matching M is the total latency of all the clients

$$\text{cost}(M) = \sum_{u \in U} \lambda_{M(u)}(d(M(u))) = \sum_{v \in V} d(v) \cdot \lambda_v(d(v)). \tag{1}$$

The last equality expresses that each of the $d(v)$ clients matched to v suffers latency $\lambda_v(d(v))$.

Our server matching game is a form of the *congestion games* studied by Rosenthal [21], who showed that every congestion game has a pure strategy Nash equilibrium. A strategy in our server matching game is the choice of the servers. We focus on the *atomic* version of the game, where each client chooses exactly one server.

1.2   Overview of Results

This paper provides an overview of our ongoing research on the server matching game. Our primary goal here is to introduce the problem to the algorithms community, recap our main results, and suggest some useful research directions. Along the way, we extend our earlier results in a couple of interesting directions.

In Section 2, we analyze the price of anarchy. Under linear latency functions, we show that the cost of a worst-case Nash solution is at most 2.5 times the optimal. (In this paper, we extend the previous proof so that the bound holds for the latency functions of the form $\lambda_v(x) = a_v x + b_v$.) We also present a lower bound construction showing that a worst-case Nash is at least 2.01 times worse than optimal. If the latency functions are degree p monomials, then the price of anarchy, as measured using the L_p norm, is $(p/\log p)(1 + o(1))$.

In Section 3, we consider bounds on individual client's latency. In particular, if a server has degree $d_{\text{opt}}(v)$ in an optimal solution, then its degree in worst-case Nash is $\Theta(d_{\text{opt}}(v) + \log m / \log\log m)$. This bounds the *individual latency* for any user in the worst-case Nash compared to the optimal.

Finally, we discuss some algorithmic questions concerning the computation of optimal, Nash, and greedy matchings. We conclude the paper with some open problems and suggestions for research directions.

1.3   Related Work

Koutsoupias and Papadimitriou [16] considered a transportation problem where n independent agents wish to route their traffic through a network of m parallel edges; see also Czumaj and Vöcking [6, 5] and Fotakis et al. [11]. Roughgarden and Tardos [23] generalize the selfish routing to arbitrary networks. The routing games are similar to server matching, but there are important differences. In [6, 16], an agent can send traffic on any link; in our case, this is constrained by the bipartite graph. In [23], the network topology is arbitrary, but the flows are non-atomic—they can be split across multiple paths.

In the *coordinated* setting, client-server matching problems have been studied for a long time. The focus of that work has been on balancing the load across servers. However, the two problems are related because the total latency of all the clients has close connection to the total server load: in particular, when the latency function has the form x^{p-1}, the total cost of a matching is related to the L_p norm load balancing [2]. The greedy assignment of [2] is known to be within a constant factor of the optimal, but it is easy to see that greedy matchings are in general not at equilibrium. In fact, our results show that the price of anarchy is strictly smaller than the competitive ratio of the greedy, suggesting that rationality and strategy more than compensate for the lack of coordination.

Azar et al. [3] investigate the competitiveness of the greedy assignment for the maximum latency. They show that the worst-case degree of a single server is $\Theta(\log n)$ times the optimal degree. We show that a server's degree in a Nash solution increases by at most $O(d_{\mathrm{opt}} + \log m / \log \log m)$ compared to the optimal.

In the algorithmic context, Harvey et al. [14] explore the characterizations of the social optimum and propose an algorithm for computing it. Evan-Dar et al [8] study the convergence time to a Nash equilibrium matching, but their model assumes a central controller who has full knowledge of the system and can block users from switching their servers.

2   Price of Anarchy

In this section, we prove upper and lower bounds on the price of anarchy. Given an instance of the server matching problem, let M_{opt} denote an assignment realizing the social optimum, and let M_{nash} denote a Nash assignment. For server j, let O_j and N_j denote the *set of clients* assigned to j in M_{opt} and M_{nash}, respectively. We use the shorthand notation $o_j = |O_j|$ and $n_j = |N_j|$ for the cardinalities of these sets. The following lemma notes a simple but crucial condition imposed by a Nash equilibrium. Essentially it says that, in a Nash solution, a client cannot improve its latency by switching to any other server, including the server it is matched to in the optimal.

Lemma 1 (Nash Condition). *Given an optimal assignment M_{opt} and a Nash assignment M_{nash}, the following inequality holds for any two servers j, k:*

$$\lambda_j(n_j) \leq \lambda_k(n_k + 1) \quad if \quad N_j \cap O_k \neq \emptyset. \tag{2}$$

Proof. If $k = j$ or $N_j \cap O_k = \emptyset$, then the argument is trivially true, so assume that $k \neq j$ and $N_j \cap O_k \neq \emptyset$. Pick an arbitrary client $i \in N_j \cap O_k$. This client has latency $\lambda_j(n_j)$ in the Nash assignment M_{nash}. The server k is also permissible for i because $i \in O_k$. By switching to k, the client i can achieve latency $\lambda_k(n_k + 1)$. By the equilibrium property, this latency is the same or worse than its latency in M_{nash}. Thus, $\lambda_j(n_j) \leq \lambda_k(n_k + 1)$.   □

This simple Nash Condition leads to the following important inequality.

Lemma 2 (Nash Inequality). *Given an optimal assignment M_{opt} and a Nash assignment M_{nash}, the following inequality holds:*

$$\sum_{j=1}^{m} n_j \cdot \lambda_j(n_j) \ \leq \ \sum_{j=1}^{m} o_j \cdot \lambda_j(n_j + 1). \tag{3}$$

Proof. Consider the Nash Condition (2) for every client $u \in U$ with the servers $j_u = M_{\mathrm{nash}}(u)$ and $k_u = M_{\mathrm{opt}}(u)$. Summing these inequalities over all n clients, we obtain the Nash Inequality (3). $\qquad\square$

2.1   Upper Bounds

We begin with the simplest of the latency functions. In [24], we have shown that if all servers have identical linear latency function $\lambda_v(x) = x$, then the price of anarchy, namely, the worst-case ratio $\mathrm{cost}(M_{\mathrm{nash}})/\mathrm{cost}(M_{\mathrm{opt}})$ is at most $1 + 2/\sqrt{3} \approx 2.15$. For non-identical latency functions of the form $\lambda_v(x) = a_v x$, the price of anarchy is at most 2.5. In this paper, we extend that proof technique to the most general form of linear latency functions: $\lambda_v(x) = a_v x + b_v$, where a_v and b_v are non-negative constants. The parameter a_v can be interpreted as the inverse speed of server v, and b_v can be thought of as *network latency*, which depends on the (static) network topology.

Theorem 1. *Assume that every server v has a linear latency function of the form $\lambda_v(x) = a_v x + b_v$, where $a_v, b_v \geq 0$. Then the price of anarchy is at most 2.5, i.e., $\mathrm{cost}(M_{\mathrm{nash}}) \leq 2.5\,\mathrm{cost}(M_{\mathrm{opt}})$.*

Proof. The cost of an optimal assignment M_{opt} and a Nash assignment M_{nash} can be expressed as follows (recall that n_j and o_j are the cardinalities of the sets of clients assigned to server j in M_{nash} and M_{opt}, respectively).

$$\mathrm{cost}(M_{\mathrm{nash}}) = \sum_{1 \leq j \leq m} n_j(a_j n_j + b_j) = \sum_{1 \leq j \leq m} \left(a_j n_j^2 + b_j n_j\right),$$

$$\mathrm{cost}(M_{\mathrm{opt}}) = \sum_{1 \leq j \leq m} o_j \left(a_j o_j + b_j\right) = \sum_{1 \leq j \leq m} \left(a_j o_j^2 + b_j o_j\right).$$

The Nash Inequality (3) implies that:

$$\mathrm{cost}(M_{\mathrm{nash}}) = \sum_{j=1}^{m} \left(a_j n_j^2 + b_j n_j\right) \leq \sum_{j=1}^{m} o_j \lambda_j(n_j+1) = \sum_{j=1}^{m} (a_j n_j o_j + a_j o_j + b_j o_j).$$

Fix an index j, it is straightforward to verify the following equality:

$$o_j n_j = \frac{1}{3} n_j^2 + \frac{3}{4} o_j^2 - \frac{1}{3}(n_j - \frac{3}{2} o_j)^2.$$

The Nash Inequality (3) together with the above inequality implies the following:

$$\sum_{j=1}^{m} (a_j n_j^2 + b_j n_j) \le \sum_{j=1}^{m} \left(a_j \left(\frac{1}{3}n_j^2 + \frac{3}{4}o_j^2 - \frac{1}{3}(n_j - \frac{3}{2}o_j)^2 \right) + a_j o_j + b_j o_j \right),$$

$$\sum_{j=1}^{m} \left(a_j n_j^2 + \frac{3}{2}b_j n_j \right) \le \sum_{j=1}^{m} \left(\frac{9}{8}a_j o_j^2 - \frac{1}{2}a_j (n_j - \frac{3}{2}o_j)^2 + \frac{3}{2}a_j o_j + \frac{3}{2}b_j o_j \right),$$

$$\sum_{j=1}^{m} (a_j n_j^2 + b_j n_j) \le \sum_{j=1}^{m} \left(a_j \left(\frac{9}{8}o_j^2 + \frac{3}{2}o_j - \frac{1}{2}(n_j - \frac{3}{2}o_j)^2 \right) + \frac{5}{2}b_j o_j \right).$$

In order to prove $\text{cost}(M_{\text{nash}}) \le (5/2)\text{cost}(M_{\text{opt}})$, thus it suffices to prove that

$$\frac{9}{8}o_j^2 + \frac{3}{2}o_j - \frac{1}{2}(n_j - \frac{3}{2}o_j)^2 \le \frac{5}{2}o_j^2, \quad \forall\, 1 \le j \le m,$$

which is equivalent to the following simplified form

$$o_j(3 - \frac{11}{4}o_j) \le (n_j - \frac{3}{2}o_j)^2.$$

This inequality holds trivially if either $o_j = 0$ or $o_j \ge 2$. Since o_j is an integer, the only remaining case is $o_j = 1$, and in this case it is equivalent to $1/4 \le (n_j - 3/2)^2$. Because n_j is an integer, as well, this holds true and our proof is complete. $\square$

We can also bound the performance ratio of Nash assignment in terms of n (the number of clients), m (the number of servers), and server speeds. Specifically, as the number of users grows relative to the number of servers and server speeds are relatively bounded, the Nash solution tends to the optimum. We formalize this statement in the following theorem [24]:

Theorem 2. *With linear latency functions ($\lambda_v(x) = a_v x$), the price of anarchy is $1 + o(1)$ given that $n \gg m$ and server speeds are relatively bounded. Formally, the following is true:*

$$\frac{\text{cost}(M_{\text{nash}})}{\text{cost}(M_{\text{opt}})} \le 1 + \frac{m + \sqrt{\sum_{j=1}^{m} a_j \sum_{j=1}^{m} \frac{1}{a_j}}}{2n} \le 1 + \left(1 + \sqrt{\max_{1 \le j, k \le m} \frac{a_j}{a_k}} \right) \frac{m}{2n}.$$

2.2   Lower Bounds

If the latency functions are arbitrary monomials ($\lambda_v(x) = x^p$, $p \in \mathbb{N}$), then a Nash solution can be arbitrarily worse than the optimum. This holds even if all servers have identical latency functions. For instance, consider the example shown in Figure 1(a), and choose the latency function $\lambda(x) = x^p$ for all servers. In an optimal solution, $\text{cost}(M_{\text{opt}}) = 3$, and each client has latency 1. In a worst-case Nash solution (Figure 1(c)), $\text{cost}(M_{\text{nash}}) = 1 + 2^{p+1}$ and both u_2 and u_3 have latency 2^p.

We have shown in [24] that even with identical and linear latency functions, the Nash matching can be at least 2.01 times as costly as the optimal.

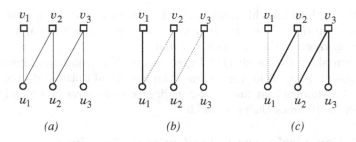

Fig. 1. (a) An instance. (b) An optimal matching, of cost 3. (c) A Nash but not optimal matching, of cost 5

Theorem 3. *In the worst case, the following lower bound holds for the price of anarchy:*

$$\frac{\mathrm{cost}(M_{\mathrm{nash}})}{\mathrm{cost}(M_{\mathrm{opt}})} > 2.01.$$

2.3   Price of Anarchy in the L_p-Norm

We have seen that the price of anarchy can be arbitrarily bad if the latency functions are arbitrary degree monomials. Let's study fixed degree monomial latency functions, and use L_p norm to measure the cost of an assignment. For any constant $p \geq 1$, we assume that a server with load ℓ and speed $1/a$ has latency $a\ell^{p-1}$. Thus, each of the ℓ_j clients matched with server j incurs latency $\lambda_j = a_j\ell_j^{p-1}$. The L_p norm measure of the total latency is $(\sum_{i=1}^{n} \lambda_i)^{1/p} = (\sum_{j=1}^{m} a_j\ell_j^{p})^{1/p}$. The case $p = 1$ is the extreme case where a server's latency is *independent* of its load—in such a case, all Nash equilibria are optimal. Thus, the interesting cases are only when $p > 1$. Our result in [24] shows that the price of anarchy with this latency measure is $(p/\log p)(1 + o(1))$.

Theorem 4. *With the L_p norm latency measure and arbitrary server speeds, the price of anarchy is bounded by* $\mathrm{cost}(M_{\mathrm{nash}})/\mathrm{cost}(M_{\mathrm{opt}}) \leq (p/\log p)(1 + o(1))$.

3   Bounds on Individual Degree

So far, we have considered only the *total* cost of a matching. We now study how well off individual users are in a worst-case Nash matching compared to the optimum. All through this section, we assume the simplest latency functions $\lambda_v(x) = x$. It is easy to extend our results to the case where servers have identical latency function $\lambda(x)$ and $x\lambda(x)$ is convex.

In general, there can be many optimal matchings, and so to which optimum should we compare the Nash matching? We prove a characterization theorem about optimal matchings, which shows that all optimal matchings are essentially the same. That is, we can speak of a *canonical optimum*. Specifically, we show that in any optimal matching, any server's degree can vary by at most one across all optimal solutions. Thus, the individual payoffs in any optimal are essentially the same as those in the canonical optimum.

We then show that a client's degree in any Nash matching increases to at most $\Theta(d_P(v) + \log m/\log\log m)$ from the canonical optimal degree $d_P(v)$. By contrast, it is known that the worst-case degree of a server in a greedy matching can be $\Theta(\log n)$ *times* the optimal [3].

3.1   A Canonical Structure of Optimal Matchings

We begin by describing the canonical structure of optimal matchings. Recall the classical notion of an *alternating path* in a graph with respect to a matching. Given an input bipartite graph $G = (U, V, E)$ and a server matching M, a (directed) *alternating path* between two servers is a path that starts with an edge in M and then alternates between edges in $E \setminus M$ and those in M. Given a server matching M, we define a partition P_M of the set of servers V in the following manner. Let d be the maximum server degree in M. Let the group V_d contain all the servers with degree d *and* all the servers reachable from a degree d server by an alternating path. Repeat this procedure for $V \setminus V_d$ until the maximum degree is zero. Thus we obtain a partition of V into at most d groups: $V = V_1 \cup V_2 \cup \ldots \cup V_d$. [3] The partition of V, in turn, corresponds to a partition of the clients $U = U_1 \cup U_2 \cup \ldots \cup U_d$ such that U_i accommodates the clients assigned by M to a server in V_i. See Fig. 2 for an illustration.

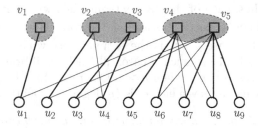

Fig. 2. An example server matching and the corresponding server partition. The three dashed ovals denote the three server groups

For a server v and a partition P, let $d_P(v)$ denote the maximum degree among all servers in v's group. We can now state some properties for partitions of optimal matchings:

Lemma 3. *If M is an optimal matching, then its partition P satisfies the following three conditions:*

---

[3] Note that $V_0 = \emptyset$, since every zero degree server is adjacent in G to at least one client, and so it is reachable by an alternating path from a server whose degree is non-zero.

1. [Bounded-Degree] If $v \in V_i$, then $i - 1 \leq d_M(v) \leq i$.
2. [Alternating-Path] If there is a server $w \in V_i$ with $d_M(w) = i-1$, then there is a server $v \in V_i$ with $d_M(v) = i$ and w is reachable from v by an alternating path.
3. [No-Back-Edge] If M assigns a client u to a server in V_i then every admissible server of u is in $\bigcup_{j=i}^{d} V_j$. (That is, there is no edge from u to any server in $V_1, V_2, \ldots, V_{i-1}$.)   □

The proof of Lemma 3 easily follows from the definition of the partition and the optimality, and it is omitted from this extended abstract. We are ready to state our result concerning the structure of optimal matchings.

Theorem 5. *All optimal matchings induce the same (thus canonical) partition of servers. If we denote by $d_P(v)$ the canonical degree of server v, defined as the maximum degree of any node in v's group in the partition, then $d_P(v) - 1 \leq d_{\mathrm{opt}}(v) \leq d_P(v)$, where $d_{\mathrm{opt}}(v)$ is the degree of v in an optimal matching.*

Proof. Consider two optimal matchings A and B, and the corresponding partitions P_A and P_B. If P_A and P_B are different, then we can pick the minimum index i such that either $V_i^A \neq V_i^B$ or $U_i^A \neq U_i^B$. Let $Z = V_i^A \setminus V_i^B$ and $R = U_i^A \setminus U_i^B$. By symmetry of A and B, we can assume that $Z \neq \emptyset$ or $R \neq \emptyset$.

By condition No-Back-Edge applied to the optimal partition P_A, clients of U_i^A can be adjacent to servers in V_i^A but are not adjacent to any server in $\bigcup_{j=1}^{i-1} V_j^A$. By condition No-Back-Edge applied to P_B, no client in $U_i^A \cap U_i^B$ is adjacent to any server in Z; and by Alternating-Path, no client in R is adjacent to servers in $V_i^A \cap V_i^B$. Thus, clients in R are the only clients that A matched to servers in Z; and both A and B match all clients of R to servers in Z.

By condition Bounded-Degree, A matches $i-1$ or i clients of R to every server in Z. But it follows from condition Alternating-Path that A matches i clients of R to at least one server of Z. Therefore, $(i-1)|R| < |Z|$. This implies that B matches at least i clients of R to a server of Z, a contradiction! We conclude that $Z = \emptyset$, $R = \emptyset$, and $P_A = P_B$.   □

3.2   Upper and Lower Bounds

In order to bound the degree ratio of a server in a Nash and an optimal matching, we introduce a transition digraph, which is a useful tool for tracking the migration of clients between two matchings.

Definition 1. *Given two server matchings M_1 and M_2, their transition digraph, denoted by $D(M_1, M_2)$, is a directed graph on the vertex set V. For each client $u \in U$, if u is matched to v_1 in M_1 and to v_2, $v_1 \neq v_2$, in M_2, then we put a directed edge $v_1 v_2$ in $D(M_1, M_2)$ (multiple edges are possible).*

We want to compare a Nash solution with an optimal solution, but it is not obvious *which* optimal matching is best to consider. The following technical lemma states that if we choose the optimal matching such that the symmetric difference $M \oplus M_{\mathrm{opt}}$ is minimal, then $D(M, M_{\mathrm{opt}})$ has nice properties.

Lemma 4. *(Harvey et al. [14]) For any server matching M, there exists an optimal matching M_{opt} such that $D(M, M_{\mathrm{opt}})$ is cycle-free and $d_{\mathrm{opt}}(v)$ is monotone decreasing along every directed path in $D(M, M_{\mathrm{opt}})$.*

The following theorem states our main result about the individual latency.

Theorem 6. *Let $d_P(v)$ be the canonical degree of a server v, and let $d_{\mathrm{nash}}(v)$ be the worst-case degree of v in a Nash matching. Then, the following bound holds:*

$$d_{\mathrm{nash}}(v) \;=\; O\left(d_P(v) + \frac{\log m}{\log \log m}\right).$$

Proof. Consider an instance of the server matching problem and a Nash solution M_{nash}. Let M_{opt} be an optimal matching such that the digraph $D(M_{\mathrm{nash}}, M_{\mathrm{opt}})$ does not have cycles and d_{opt} decreases along every directed path (Lemma 4). Let v_0 be a server where $d_{\mathrm{nash}}(v_0) - d_{\mathrm{opt}}(v_0)$ is maximal. For brevity, we put $k = d_{\mathrm{nash}}(v_0)$ and $d = d_{\mathrm{opt}}(v_0)$.

For every server v reachable on a directed path from v_0, we have $d_{\mathrm{opt}}(v) \leq d_{\mathrm{opt}}(v_0) = d$ by the choice of M_{opt}. We also have $d_{\mathrm{nash}}(v) \leq d_{\mathrm{nash}}(v_0) = k$ by the choice of v_0. Consider the vertices reachable on a directed path from v_0: Let V_ℓ be the set of servers whose Nash degree is at least ℓ for $\ell = 0, 1, \ldots, k$. Notice that an edge $v_i v_j$ in $D(M_{\mathrm{nash}}, M_{\mathrm{opt}})$ implies that v_i and v_j are permissible for a common client and by the Nash condition, we have $d_{\mathrm{nash}}(v_j) + 1 \geq d_{\mathrm{nash}}(v_i)$. Therefore if $v_i v_j \in D(M_{\mathrm{nash}}, M_{\mathrm{opt}})$ and $v_i \in V_\ell$, then $v_j \in V_{\ell-1}$. That is, every directed edge starting in V_ℓ ends in $V_{\ell-1}$.

In the digraph $D(M_{\mathrm{nash}}, M_{\mathrm{opt}})$, every server v has at least $d_{\mathrm{nash}}(v) - d_{\mathrm{opt}}(v)$ outgoing edges but at most $d_{\mathrm{opt}}(v) \leq d$ incoming edges. The set of servers V_ℓ has at least $|V_\ell| \cdot (\ell - d)$ outgoing edges, all pointing into $V_{\ell-1}$, but $V_{\ell-1}$ has at most $d|V_{\ell-1}|$ incoming edges. This implies that

$$d|V_{\ell-1}| \;\geq\; (\ell - d)|V_\ell| \quad \Leftrightarrow \quad |V_{\ell-1}| \;\geq\; \frac{\ell - d}{d}|V_\ell|, \quad \forall\, \ell = 1, \ldots, k.$$

By the Stirling formula and using the fact that $|V_k| \geq 1$, we conclude that

$$m \geq |V_d| \;\geq\; \frac{1}{d}|V_{d+1}| \;\geq\; \frac{2!}{d^2}|V_{d+2}| \;\geq\; \cdots$$

$$\geq \frac{(k-d)!}{d^{k-d}}|V_k| \;\geq\; \frac{(k-d)!}{d^{k-d}} \;=\; \Omega\left(\left(\frac{k-d}{ed}\right)^{k-d}\right).$$

So $k = O(\log m / \log \log m)$ if $k > 3d$, and $k = O(d)$ otherwise. □

This bound is the best possible, as the lower bound construction shows below.

Theorem 7. *Assuming identical and linear latency functions, there is an instance of server matching where a client v has the following property:*

$$d_{\mathrm{opt}}(v) = 1 \quad \text{and} \quad d_{\mathrm{nash}}(v) = \Theta\left(\frac{\log m}{\log \log m}\right).$$

Proof. Consider n clients and $m = n$ servers. In the input graph G, one client is paired with one unique server. We partition the n client-server pairs into $k + 2$ classes $V_0, V_1, \ldots, V_{k+1}$ such that $|V_{k+1}| = 1$, $|V_k| = k$, and $|V_\ell| = k \cdot k!/\ell!$ for $\ell = 0, \ldots, k - 1$. We assume that $n = 1 + k + \sum_{\ell=0}^{k-1} k \cdot k!/\ell!$ for some $k \in \mathbb{N}$, therefore $n = \Theta(k \cdot k!)$ and $k = \Theta(\log n / \log \log n)$. The server $v \in V_{k+1}$ is adjacent to the unique client in V_{k+1} and to all k clients in V_k. Every server in V_ℓ, $\ell = 1, \ldots, k$, is adjacent to one client in V_ℓ and ℓ clients in $V_{\ell-1}$. (See Fig. 3 for the case $k = 2$.)

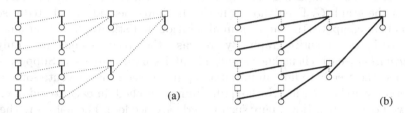

Fig. 3. Our construction for $k = 2$: (a) The optimal matching, and (b) The worst case Nash matching

The optimal matching M_{opt} is a one-to-one matching between the clients and the servers. We denote by U_ℓ the set of clients matched to V_ℓ in M_{opt}. Now we describe a Nash matching, M_{nash}, where every client in U_ℓ, $\ell = 0, \ldots, k$, is matched to an adjacent server in $V_{\ell+1}$ (and the client in V_{k+1} is matched to the server in V_{k+1}). We show that M_{nash} is Nash. Indeed, for every server $v_\ell \in V_\ell$, $\ell = 0, \ldots, k + 1$, we have $d_{\mathrm{nash}}(v) = \ell$. So no client in V_ℓ can improve its own payoff by switching from a server in $V_{\ell+1}$ to one in V_ℓ. The client in V_{k+1} suffers latency 1 at M_{opt} but its latency jumps to $k = \Theta(\log m / \log \log m)$ in M_{nash}. □

4   Algorithmic Issues

Nash equilibrium is an appealing concept in decentralized systems with selfish players. Unfortunately, they are often difficult to compute. Indeed, computing Nash equilibria in general games is considered one of the outstanding complexity questions [19]. Because of this difficulty, we are motivated to consider a simple *greedy* scheme for matching clients to servers: *every client (in order of arrival) chooses a permissible server with the minimum load, and this choice is permanent.* The order of clients is arbitrary, and may be determined by an adversary. Since the clients choose servers sequentially, the resulting greedy matching depends on this order.

We have analyzed the worst-case ratio between a greedy and an optimal matching for linear latency functions, and shown that this ratio is bounded [24].

Theorem 8. *Assuming linear latency functions of the form $\lambda_v(x) = a_v x$, the greedy matching has cost at most $17/3 \approx 5.33$ times the optimal.*

If, in addition, the server speeds are relatively bounded and $n \gg m$, then the upper bound improves to $4 + o(1)$.

4.1   Convergence to Nash Matchings

An optimal matching can be computed in polynomial time (e.g., using min-max flow [7] or the Hungarian method [17]). In particular, Harvey et al. [14] has recently proposed an $O(n \cdot |E|)$ time algorithm, where $|E|$ denotes the number of edges of the bipartite graph G.

One serious drawback of this algorithm is that it requires complete knowledge about the graph G. Even though it can be implemented distributedly so that each client computes the same optimal solution, but still every one of them would have to know the entire graph. By contrast, the greedy scheme uses only *local information*—each client knows only about its permissible set. Suppose clients first use the greedy to find an initial assignment; then they execute some rounds of server switches, until a Nash equilibrium is reached; in each round, a client is allowed one switch. How many such rounds are needed? The following theorems give partial answer to this question. (A switch is strict if it strictly improves a client's latency; it is non-strict if the latency either improves or stays the same.)

Theorem 9. *Assuming identical linear latency functions of the form $\lambda_v(x) = x$, there are instances that require $\Omega(m)$ rounds of strict switches before a Nash equilibrium is reached. Starting with any assignment, $O(n^2)$ strict switches suffice to reach a Nash equilibrium.*

Proof. We describe a construction of m servers and $n = (m^2 + m)/2 - 1$ clients such that if we allow the clients switch in a prescribed order then they reach a Nash equilibrium in m rounds. For every server v_i, $i = 1, 2, \ldots, m - 1$, there is a client $u_{j(i)}$ whose permissible servers are v_i and v_{i+1}. Let $j(1) = 1$ and $j(i) = 1 + \sum_{\ell=0}^{i-1}(m-\ell)$. For every $i = 1, 2, \ldots, m-1$, the clients $u_{j(i)+1}, u_{j(i)+2}, \ldots, u_{j(i+1)-1}$ are all connected to a unique permissible server v_i. These clients are assigned to v_i in every matching (see an example in Fig. 4.1).

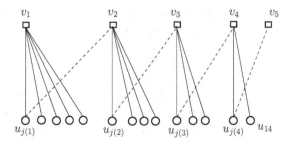

Fig. 4. Our construction for $m = 5$. Solid lines indicate a greedy matching

In a greedy solution M_{greedy}, every $u_{j(i)}$, $i = 1, 2, \ldots, m - 1$, is assigned to v_i. In this matching, only $u_{j(m-1)}$ can improve its latency by switching from

v_{m-1} to v_m. Once $u_{j(m-1)}$ has swapped, the degree of v_{m-1} drops by one, and now $u_{j(m-2)}$ has an incentive to swap from v_{m-2} to v_{m-1}. In round i, $i = 1, 2, \ldots m - 1$, client $u_{j(m-i)}$ switches from v_{m-i} to v_{m-i+1}. Therefore, it takes $m - 1$ rounds until they reach a Nash equilibrium matching.

On the other hand, every strict switch strictly decreases the total cost. Since the total cost can have at most $O(n^2)$ different values, this also an upper bound bounds of number of strict switches. □

Theorem 10. *Assuming identical linear latency functions ($\lambda_v(x) = x$), starting with an arbitrary assignment, one can reach an optimal (best-case Nash) assignment in at most n non-strict switches.*

Proof. Consider an arbitrary matching M and an optimal matching M_{opt} as in Lemma 4. We show that M can be transformed into M_{opt} with a sequence of at most n switches, none of which increases the the total cost. It is enough to show that there is a directed edge $v_i v_j$ in $D(M, M_{\mathrm{opt}})$ such that $d_M(v_i) \geq d_M(v_j)$, because then the client corresponding to $v_i v_j$ can switch from v_j to v_i without increasing the total cost, and this eliminates the edge $v_i v_j$ from $D(M, M_{\mathrm{opt}})$. By swapping clients along all directed edges of $D(M, M_{\mathrm{opt}})$, we transform the matching M into M_{opt}. The transformation terminates in n switches, since $D(M, M_{\mathrm{opt}})$ has at most one edge for every client.

Suppose, to the contrary, that $d_M(v_i) < d_M(v_j)$ along every directed edge $v_i v_j$ of $D(M, M_{\mathrm{opt}})$. By Lemma 4, $d_{\mathrm{opt}}(v_i) \geq d_{\mathrm{opt}}(v_j)$ along every edge $v_i v_j$ and $D(M, M_{\mathrm{opt}})$ contains no cycle. At every server v with no outgoing edges, $d_M(v) < d_{\mathrm{opt}}(v)$ by the definition of $D(M, M_{\mathrm{opt}})$. Therefore, $d_M(v) < d_{\mathrm{opt}}(v)$ for every server incident to edges of D and $d_M(v) = d_{\mathrm{opt}}(v)$ for all other servers. A contradiction, since $\sum_{v \in V} d_M(v) = \sum_{v \in V} d_{\mathrm{opt}}(v) = n$. □

5   Future Directions

The problem of assigning clients (jobs) to servers (machines) dates back to the earliest days of distributed computing or scheduling. However, the decentralized nature of the Internet has added a new, but fundamental, twist to this classical problem—Internet-centric applications like P2P, online gaming, grid computing and other web services lack a coordinating authority and users are selfish agents who are free to choose to maximize their own utility. In such a selfish world, (centralized) social optimum is frequently unstable, and one has little choice but to embrace the anarchy of selfish users. Our work raises as many questions as it settles. In the following, we propose some natural directions for future work.

● **Best-Case Nash.** When analyzing the worst-case Nash, we have followed the traditional TCS approach. It would be interesting to approach the problem from the optimistic side too, and study the best-case Nash. With linear latency functions, perhaps the price of anarchy is modest enough to not worry about the best case. However, with higher order latency functions, the worst-case Nash

may be too unattractive a solution, and it would be interesting to obtain bounds
for the best-case Nash.

• **Coping with a Dynamic Client Set.** We have assumed a static client
group. In practice, new clients constantly arrive and old ones leave. Little is
know about the loss of efficiency in such a *dynamic* setting. One basic problem
is to investigate the price of anarchy where we compare centralized optimum to
a solution in which clients are always at Nash—that is, whenever a new client
arrives or an old one leaves, the remaining set recomputes a Nash matching.

• **Truthfulness and Mechanism Design.** Because players are assumed to
be selfish, and there is no central authority, how can one ensure that they are
truthful? In our context, what incentive does a server have to truthfully declare
its load? Nisan and Ronen [18] have advocated using the VCG (Vickrey-Clarke-
Grove) mechanism to compensate servers in a scheduling application. Ours is
another natural setting where further research of this kind is needed.

• **From Greedy to Nash.** The matching determined by the online greedy
scheme is within a constant factor of the social optimum, but it may not be
stable—some of the users may want to switch to a different server. It may be
interesting to investigate how to transform a greedy matching into a Nash.

References

1. E. Anshelevich, A. Dasgupta, E. Tardos, and T. Wexler. Near-optimal network
   design with selfish agents. In *Proc. 35th STOC*, ACM Press, 2003, pp. 511–520.
2. A. Awerbuch, A. Azar, E. F. Grove, P. Krishnan, M. Y. Kao, and J. S. Vitter.
   Load balancing in the L_p norm. In *Proc. 36th FOCS*, 1995, IEEE, pp. 383–391.
3. Y. Azar, J. Naor, and R. Rom. The competitiveness of online assignments. *Journal
   of Algorithms* **18** (2) (1995), 221–237.
4. Berkeley open infrastructure for network comp., http://boinc.ssl.berkely.edu.
5. A. Czumaj, P. Krysta, and B. Vöcking. Selfish traffic allocation for server farms.
   In *Proc. 34th SOTC*, ACM Press, 2002, pp. 287–296.
6. A. Czumaj and B. Vöcking. Tight bounds for worst-case equilibria. In *Proc. of
   13th ACM-SIAM Sympos. Discrete Algorithms*, ACM Press, 2002, pp. 413–420.
7. J. Edmonds and R.M. Karp. Theoretical improvement in algorithmic efficiency for
   network flow problems. *Journal of the ACM* **19** (2) (1972), 248–264.
8. E. Even-Dar, A. Kesselman, and Y. Mansour. Convergence time to Nash equilibria.
   In *Proc. 30th ICALP*, vol. 2719 of LNCS, Springer-Verlag, 2003, pp. 502-513.
9. Everquest, http://everquest.station.sony.com.
10. A. Fabrikant, A. Luthra, E. Maneva, C.H. Papadimitriou, and S. Shenker. On a
    network creation game. In *Proc. 22nd PODC*, ACM Press, 2003, pp. 347–351.
11. D. Fotakis, S.C. Kontogiannis, E. Koutsoupias, M. Mavronicolas, and P.G. Spi-
    rakis. The structure and complexity of Nash equilibria for a selfish routing game.
    In *Proc. 29th ICALP*, vol. 2380 of LNCS, Springer-Verlag, 2002, pp. 123–134.
12. Freenet, http://freenetproject.org.
13. Gnutella, http://www.gnutella.com.
14. N. J.A. Harvey, R. E. Ladner, L. Lovász, and T. Tamir. Semi-matchings for bipar-
    tite graphs and load balancing. In *Proc. 8th WADS*, vol. 2748 of LNCS, Springer-
    Verlag, 2003, pp. 294–306.
15. KaZaa, http://www.kazaa.com.

16. E. Koutsoupias and C. Papadimitriou. Worst-case equilibria. In *Proc. 16th STACS*, vol. 1563 of LNCS, Springer-Verlag, 1999, pp. 404-413.
17. H. W. Kuhn. The Hungarian method for the assignment problem. *Naval Research Logistics Quarterly* **2** (1955), 83–97.
18. N. Nisan and A. Ronen. Computationally feasible VCG mechanisms. *Proc. 2nd Conf. on EC*, 2000, ACM Press, pp. 242–252.
19. C. Papadimitriou. Algorithms, games, and the internet. In *Proc. 33rd Sympos. on Theory of Computing*, ACM Press, 2001, pp. 749-753.
20. PlanetLab, http://www.planet-lab.org.
21. R. W. Rosenthal. A class of games possessing pure-strategy Nash equilibria. *International Journal of Game Theory* **2** (1973), 65–67.
22. T. Roughgarden. *Selfish routing*. PhD thesis, Cornell University, 2002.
23. T. Roughgarden and E. Tardos. How bad is selfish routing? *Journal of the ACM* **49** (2002), 235–259.
24. S. Suri, C.D. Tóth, and Y. Zhou. Selfish load balancing and atomic congestion games. In *Proc. 16th SPAA*, ACM Press, 2004, pp. 188–195.

Bandwidth Allocation in Networks: A Single Dual Update Subroutine for Multiple Objectives

Sung-woo Cho[1,*] and Ashish Goel[2,**]

[1] Department of Computer Science, University of Southern California
sungwooc@cs.usc.edu
[2] Department of Management Science and Engineering and
(by courtesy) Computer Science, Stanford University
ashishg@stanford.edu

Abstract. We study the bandwidth allocation problem

Maximize $U(x)$, subject to
$$Ax \leq c; x \geq 0$$

where U is a utility function, x is a bandwidth allocation vector, and $Ax \leq c$ represent the capacity constraints. We consider the class of *canonical* utility functions, consisting of functions U that are symmetric, non-decreasing, concave, and satisfy $U(0) = 0$. We present a single dual update subroutine that results in a primal solution which is a logarithmic approximation, simultaneously, for all canonical utility functions. The dual update subroutine lends itself to an efficient distributed implementation.

We then employ the fractional packing framework to prove that at most $O(m \log m)$ iterations of the dual update subroutine are required; here m is the number of edges in the network.

1 Introduction

In this paper, we revisit the classic problem of distributed allocation of bandwidths to flows in a network (see [13, 3, 16, 1, 4, 2, 7] for some of the recent research on the problem). We will assume that the route for each flow is given upfront. Formally, the problem is:

Maximize $U(x)$, subject to
$$Ax \leq c; x \geq 0.$$

Here, $x = \langle x_1, x_2, \ldots, x_n \rangle$ is an allocation of bandwidths to n flows, U is an arbitrary n-variate "utility function", and $Ax \leq c$ are linear capacity constraints (the matrix A encodes the routes for each flow). Let R denote the ratio of the

---

* Research supported by NSF Award CCR-0133968.
** Research supported by NSF CAREER Award 0133968.

A. López-Ortiz and A. Hamel (Eds.): CAAN 2004, LNCS 3405, pp. 28–41, 2005.

largest capacity to the smallest capacity, and m denote the number of links in the network. We present an efficient dual update subroutine for this problem. Repeated invocations of this dual update subroutine result in a primal solution that simultaneously approximates the optimum solution for a large class of utility functions U. More specifically, our algorithm guarantees an $O(\log n + \log R)$ approximation simultaneously for all utility functions U that are symmetric, concave, non-decreasing, and satisfy $U(0) = 0$. Before presenting more details about our results and techniques, we review the related work and the motivation for our problem.

Motivation and Related Work: The Transport Control Protocol (TCP) is by far the most widely used solution to this problem is practice. In more abstract settings, Kelly, Maulloo, and Tan [13] proposed a distributed algorithm for this problem for the case where $U(x) = \sum_i U_i(x_i)$, and each U_i is a concave function. Their algorithm uses a primal-dual framework where the dual prices (which they call shadow prices) are maintained by "edge-agents" and primal flows are maintained by the "flow-agents". All communication is local, i.e., takes place between a flow-agent and an edge-agent on the path of the flow. The resulting solution is proportional with respect to the dual prices, and hence, their framework is widely referred to as the "proportional-fairness" framework. Subsequent work by Low, Peterson, and Wang [16] and others has shown that several variants of TCP essentially perform the above computation for different choices of utility functions. Since the behavior of different variants of TCP is quite different (different variants work better in different settings), the above work raises the following natural question: *Is it possible to obtain solutions which are simultaneously good for a wide range of utility functions?*.

Bartal, Byers, and Raz [3] presented a distributed algorithm for the above problem when $U(x) = \sum_i x_i$. Unlike the work of Kelly *et al.*, this work presents a running time analysis. They prove that a simple local computation along with local communication can lead to almost optimum solutions in polynomial time. Their work builds on the positive linear programming framework of Luby and Nisan [17]; for their problem, the positive linear programming framework is essentially identical to the fractional packing framework developed by Plotkin, Shmoys, and Tardos [19], and later simplified by Garg and Konemann [6]. Each edge-agent maintains dual costs, and each flow-agent uses these dual costs to update its own flow. Recently, Garg and Young [7] have shown how a simple MIMD (multiplicative increase multiplicative decrease) protocol can approximate the above objective. These results lead to the following natural question: *Can we obtain distributed algorithms with similar rigorous running time analyses for more involved utility functions?*.

Building on a series of papers about multi-objective fairness [14, 10], Kleinberg and Kumar [15] studied the problem of bandwidth allocation in a centralized setting with multiple fairness objectives. Goel and Meyerson [8] and Bhargava, Goel, and Meyerson [5] later expanded this work to a large class of linear programs and related it to simultaneous optimization [9]. In particular, they considered the class of utility functions U which are symmetric, concave, non-decreasing, and satisfy $U(0) = 0$. They presented an algorithm for computing a single allocation which is simultaneously an $O(\log n + \log R)$ approximation for all such utility functions. Their work builds on the notion of majorization due to Hardy, Littlewood, and Polya [11, 12, 18]; to obtain efficient solutions they use the fractional packing framework of Plotkin, Shmoys, and Tardos [19]. This leaves open the following question: *Can there be efficient distributed algorithms which achieve the same results?*

All three questions raised above are very similar, even though they arise in different contexts. They point towards a need for distributed bandwidth allocation algorithms which are good across multiple objectives and have provably good running times. We address precisely this problem. It is our hope that this line of research would ultimately lead to protocols which are comparable to TCP in simplicity (our algorithm is not) but perform well across multiple objectives.

The class of utility functions we consider is not arbitrarily chosen. This is a large class, and contains the important subclass $\sum_i f(x_i)$ where f is a uni-variate concave function (f must also be non-decreasing and $f(0)$ must be 0). Concavity is a natural restriction, since it corresponds to the "law of diminishing returns" from economics. Symmetry corresponds to saying that all users are equally important[1]. The requirements $U(0) = 0$ and U non-decreasing are natural in the setting of the bandwidth allocation problem. This class includes important special functions such as min, $\sum_i \log(1 + x_i)$, and $\sum_i -x_i \log x_i$ (the entropy). It also contains a series of functions which together imply max-min fairness. Most interestingly, there is a concrete connection between this class and our intuitive notion of fairness. Suppose there exists some function which measures the fairness of an allocation. It seems natural that the allocation (x_1, x_2) should be deemed as fair as (x_2, x_1) and less fair than $(\frac{x_1+x_2}{2}, \frac{x_1+x_2}{2})$. This assumption implies that for any natural definition of fairness, maximizing fairness should be equivalent to maximizing some symmetric concave function; certainly, all the definitions of fairness that we found in literature are of this form.

We will use the term *canonical* utility functions to refer to this class[2].

Our Result and a Summary of Our Techniques: We present an algorithm that simultaneously approximates all canonical utility functions U up to a factor $O(\log n + \log R)$. Our algorithm performs $O(m \log m)$ iterations of a *single* dual subroutine, i.e., the subroutine does not depend on U. Here m is the number of edges in the network. Each invocation of the subroutine requires $O(\log n)$ steps. During each invocation, a flow-agent sends $O(1)$ messages of length $O(\log n + \log R)$. Also, during each invocation, each flow-agent exchanges $O(1)$ messages with each edge-agent on its route; this exchange between flow-agents and edge-

---

[1] Notice that the constraints are not required to be symmetric, and hence, the optimum solution need not be symmetric even though the objective function is symmetric.

[2] The symmetry requirement implies that our class does not contain all utility functions to which the proportional-fairness framework applies. However, since our class does not require the utility function to be a sum of uni-variate functions, it contains important functions such as min which can not be addressed using the proportional-fairness framework. Hence the two classes of utility functions are incomparable.

agents is also used in the work of Kelly *et al.* [13] and Bartal *et al.* [3] and can be implemented efficiently by piggybacking information over data packets belonging to a flow.

To obtain the above results, we first describe (section 2) the centralized algorithm due to Goel and Meyerson [8]. This algorithm requires a solution to n fractional packing problems. Each packing problem requires a different dual optimization subroutine. A distributed implementation of this algorithm would require each edge-agent to maintain n different dual costs. Also, it is not clear how the dual subroutine can be implemented in a distributed fashion even for a single packing problem. To address these problems. we present (section 3) a single dual subroutine that approximates all the n different dual subroutines used in the centralized version. In this dual subroutine, the i-th flow-agent updates its cost by $1/C_i'$ where C_i' denotes the sum of the dual costs of all the flow-agents which are no more expensive than flow i. Because of its simplicity, this new dual subroutine is amenable to a distributed implementation. Since the same dual subroutine is used for all the fractional packing problems, the dual costs for the different packing problems grow identically, eliminating the need for maintaining multiple dual costs. Finally, we show how this dual subroutine can be used in conjunction with the fractional packing algorithm of Garg and Konemann [6] to obtain our result. Even though we use the algorithm of Garg and Konemann, we need to make several non-trivial changes to their analysis; details are presented in section 4. We prove that $O(m \log m)$ iterations suffice.

While we present our algorithm in the setting in which flows needs to be computed from scratch, it can easily be adapted to the setting where we need to improve existing flows incrementally. Even in a centralized setting, our algorithm would be considerably more efficient than earlier solutions which involve solving n fractional packings. The number of iterations made by our algorithm is identical to that of the single-objective algorithm by Bartal, Byers, and Raz [3]. Further, the guarantee on the approximation ratio for our algorithm matches (up to constant factors) the best known guarantee for simultaneous optimization even in a centralized setting (see theorem 2) and is in the same ballpark as a lower bound of $O(\log n / \log \log n)$ due to Kleinberg and Kumar [15] for this problem[3].

Future Directions: To completely answer the questions posed earlier in the introduction, we would need to have a completely distributed and efficient algorithm for simultaneous approximation of all canonical utility functions. We believe that the dual update procedure and the analysis outlined in this paper is a good step in that direction. However, we still need to address several issues before we can obtain simple and practical protocols that perform simultaneous optimization:

1. Our algorithm, while quite efficient, is not "local"; it requires a small amount of communication between different flow-agents. Consequently, it seems hard to design a simple TCP-like protocol which mimics our distributed algorithm. However, the simplicity of our dual solution makes us optimistic that a simple

---

[3] Kleinberg and Kumar studied the case where $R = 1$.

protocol which only exchanges local messages can be used to replace our distributed algorithm for computing the dual subroutine. Such a protocol would be a significant development, both theoretically as well as in terms of its practical impact.

2. Our analysis requires $m \log m$ iterations (ignoring constant factors) of the dual update procedure. Also, we can construct simple examples where any flow improvement algorithm using our dual update subroutine will take at least $\tilde{\Omega}(n)$ steps, where the $\tilde{\Omega}$ notation hides polylogarithmic factors. To be considered practical, the number of iterations must be sub-linear.

2   Background: Simultaneous Optimization in a Centralized Setting

In this section we describe the framework for simultaneous optimization of linear programs developed by Bhargava, Goel, and Meyerson and by Goel and Meyerson [5, 8]. This framework works in a centralized setting. We also point out the difficulties in making it distributed.

Suppose we are given a linear set of constraints $\{Ax \leq c, x \geq 0\}$ where $x = \langle x_1, x_2, \ldots, x_n \rangle$ is an n-dimensional vector. Recall that we call a utility function U *canonical* if U is symmetric and concave in its arguments, $U(0) = 0$, and U is non-decreasing in each argument. Define the k-th prefix, $P_k(x)$ to be the sum of the k smallest components of x (not $x_1 + x_2 + \ldots + x_k$ but $x_{\sigma(1)} + x_{\sigma(1)} + \ldots x_{\sigma(k)}$ where σ is the permutation that sorts x in increasing order.). Let P_k^* denote the maximum possible value of $P_k(x)$ subject to the given constraints

Definition 1 Approximate Majorization: *A feasible solution x to the above problem is said to be α-majorized if $P_k(x) \geq P_k(y)/\alpha$ for all $1 \leq k \leq n$ and all feasible solutions y.*

Informally, the k poorest users in an α-majorized solution get at least $1/\alpha$ times as much resource as the k poorest individuals in any other feasible allocation. Intuitively, this seems to have some ties to fairness. The following theorem [8] makes this intuition concrete:

Theorem 1. *A feasible solution x is α-majorized if and only if $U(x) \geq U(y)/\alpha$ for all feasible solutions y and all canonical utility functions U.*

In fact, the above theorem holds for an arbitrary set of constraints (integer, convex etc.) as long as they imply $x \geq 0$. Thus the notion of α-majorization captures simultaneous optimization. For this framework to be useful, we need to demonstrate that α-majorized solutions exist with small values of α; the following theorem does exactly that [8].

Theorem 2. *If the set of feasible solutions is convex and non-negative, then there exists an $O(\log \frac{P_n^*}{nP_1^*})$-majorized solution.*

For many problems of interest, the above theorem translates into $\alpha = O(\log n)$. For example, for the bandwidth allocation problem with unit capacities, $P_n^* \leq n$ whereas $P_1^* \geq 1/n$, implying the existence of an $O(\log n)$-majorized solution. For non-uniform capacities, the guarantee becomes $O(\log n + \log R)$ where R is the ratio of the maximum to the minimum capacity. However, even if there exists an α-majorized solution, it is not clear a priori that finding such a solution is computationally tractable. The next theorem [8] resolves this issue assuming linear constraints. Here, α^* is the smallest possible value of α for which an α-majorized solution exists.

Theorem 3. *Given the constraints* $\{Ax \leq c, x \geq 0\}$, *both* α^* *and an* α^*-*majorized solution can be found in time polynomial in n and the number of constraints.*

Similar techniques were developed by Tamir [20] to compute majorized elements. Goel and Meyerson [8] also demonstrate how the above framework can be applied to several integer programming problems using suitable rounding schemes.

Let us focus on a proof of theorem 3 that highlights the difficulties involved in making the above framework carry over in a distributed setting. We will restrict ourself to the bandwidth allocation problem for simplicity, where A and b are both required to be non-negative. In order to compute α^*, we first need to compute P_k^*. Computing P_k^* is equivalent to solving the following linear program:

$$\text{Minimize } \lambda_k \text{ subject to:}$$
$$Ax \leq \lambda_k c$$
$$\sum_{i \in S} x_i \geq 1 \text{ for all } S \subseteq \{1, \cdots, n\} \text{ with } |S| = k$$

P_k^* would be $1/\lambda_k^*$ where λ_k^* is the solution to the above linear program. The linear program described above is a fractional packing problem, and can be solved efficiently using the framework of Plotkin, Shmoys, and Tardos [19] or the simplification thereof by Garg and Konemann [6]. In order to solve the fractional packing problem efficiently, we need a *dual optimization subroutine* to solve the following program:

$$\beta_k(l) = \text{Maximize } P_k(x) \text{ subject to:}$$
$$Cx = 1; x \geq 0$$

where $l(e) \geq 0$ is the "dual cost" of edge e and $C_i \geq 0$ represents the dual cost for flow i. The dual cost C_i for flow i is computed by simply adding the dual costs $l(e)$ of each edge e used by the flow. It is important to note that the dual costs are artifacts of the solution methodology and do not correspond to any real entity in the original problem.

Lemma 1. *[8] For any k, there exists an optimum solution x to the dual problem such that*

1. $C_i \leq C_j \Rightarrow x_i \geq x_j$,
2. *The solution x is two-valued. In particular, there exists a value γ such that for all i, $x_i = 0$ or γ.*

We will use the term "candidate dual solutions" to refer to feasible solutions which satisfy properties 1 and 2 in the above lemma. There are only n candidate dual solutions, and they can be enumerated in time $O(n \log n)$. This leads to an efficient dual optimization subroutine, and hence, an efficient solution to the fractional packing problem (the fractional packing problem can be solved by making polynomially many invocations of the dual subroutine). In order to find P_k^* for all $k, 1 \le k \le n$, we need to solve n fractional packing problems. The quantity α^* and the corresponding α^*-majorized solution can then be computed using similar techniques.

The above discussion pertained to centralized solutions. As mentioned before, a distributed solution to the bandwidth allocation problem would be quite desirable. The following issues need to be addressed in order to make the above framework distributed:

1. We need to find a distributed solution to the dual problem. Enumerating all the candidate dual solutions is efficient in a centralized setting. But in a distributed setting, this would require communicating all the dual costs to all the users.
2. In the centralized framework, we need to solve n fractional packing problems, one for each P_k^*. The dual costs for each problem need not be the same as the algorithm for each problem progresses. Hence, the network would need to keep track of n different dual costs per edge.
3. The centralized solution first computes all P_k^* and then computes α^*. it would be desirable for a distributed algorithm to directly obtain a small α.

We address the above issues in the next section by giving a simple dual subroutine that simultaneously approximates the dual problems for each P_k^* and can be made to work efficiently in a distributed setting. Since the same dual solution is used for each of the n packing problems, the dual costs for each problem remain the same as the algorithm progresses. In section 3 we show how the framework of Garg and Konemann can be adapted to our setting where multiple primal problems are being approximated using a single dual subroutine.

3   Distributed Simultaneous Optimization of the Dual Problems

In this section we will assume that the users are re-numbered in increasing order of dual costs i.e. $C_1 \le C_2 \le \ldots \le C_n$. We first use lemma 1 to describe the j-th candidate dual solution, $x^{(j)}$:

$$x_1^{(j)} = x_2^{(j)} = \ldots = x_j^{(j)} = \frac{1}{\sum_{i=1}^j C_i}, \text{ and } x_i^{(j)} = 0 \text{ for } i > j.$$

Let C_j' denote the quantity $\sum_{i=1}^j C_i$. Let $\bar{x}$ denote the upper envelope of all the candidate dual solutions, i.e., $\bar{x}_j = 1/C_j'$. This upper envelope solution has several desirable properties. First, it dominates all the candidate dual solutions and hence, $P_k(\overline{x}) \geq \beta_k(l)$ for all $1 \leq k \leq n$. Further, the cost of $\overline{x}$ is not too much:

Lemma 2. $C\overline{x} \leq 1 + \ln n + \ln \frac{C_n}{C_1}$.

Proof.

$$C \cdot \overline{x} \leq \sum_{j=1}^{n} \frac{C_j}{C_j'}$$

$$\leq 1 + \int_{C_1}^{C_n'} (1/x)dx = 1 + \ln \frac{C_n'}{C_1}$$

$$\leq 1 + \ln \frac{nC_n}{C_1} = 1 + \ln n + \ln \frac{C_n}{C_1}$$

Recall that the dual cost for each flow is just the sum of the dual cost for all the edges used by the flow. Hence, each flow-agent can obtain an estimate of its own cost by merely sending a control packet which sums up the cost of all the edges on the route for the flow. In order to compute C_i', user i needs to know the sum of the costs of all the "cheaper" flows. This can be accomplished using a logical balanced tree T on all the flow-agents. Each node on the tree contains an array $A[1..n]$. Starting from the leaves, each node x_i records its cost in $A[i]$ and passes the array to its parent. The parent node x_j now merges the arrays which it received from its children and overwrites its cost on $A[j]$. Then it again passes the array to its parent. After $\log n$ recursive steps, the cost information of all the nodes is delivered to the root of the tree. Once the root collects the information, it distributes the array down the tree. The total number of steps needed is $2 \log n$ and each node sends at most three message. Unfortunately, the size of each message is quite large. We solve this problem by reducing the size of the array to $O(\log \frac{C_n}{C_1})$. The basic idea is to group the costs into logarithmic number of sets. This can be done by rounding C_i down to powers of 2. Let W_i be the rounded cost of agent i, i.e.

$$W_i = 2^{\lfloor \log C_i \rfloor}.$$

When agent i receives arrays from its children, it merges them, increments the entry of array $A[\log \frac{W_i}{W_1}]$ by one, and then passes the array to its parents. The rest of the procedure is same as before. Rounding allows us to keep only $O(\log \frac{C_n}{C_1})$ number of groups, thus reducing the message size while introducing only a constant factor increase in the cost of the dual solution.

The following theorem summarizes the results of this section:

Theorem 4. *Given dual costs C, our distributed algorithm finds a dual solution x such that $P_k(x) \geq \beta_k(l)$ and $Cx = O(\log n + \log \max\{C_i\}/\min\{C_i\})$. The algorithm uses $O(\log n)$ message passing steps. Each flow-agent sends at most 3 messages of size $O(\log \max\{C_i\}/\min\{C_i\})$ each and does $O(\log \max\{C_i\}/\min\{C_i\})$ amount of computation.*

Armed with this distributed algorithm for simultaneous approximation of the dual problems, we now show how the fractional packing algorithm of Garg and Konemann can be adapted to obtain an $O(\log n)$-majorized solution for the bandwidth allocation problem.

4   Simultaneous Optimization for the Primal Problems

We will now show how to use our dual subroutine from the previous section to achieve simultaneous optimization for the bandwidth allocation problems. While we will follow the structure and the notation of the fractional packing algorithm of Garg and Konemann [6], we can not use their result in a black-box fashion for several reasons. Since we are essentially solving n primal problems at the same time, we need to make sure that the same stopping condition can be used for each problem. Also, the approximation ratio for the dual problem depends on the ratio $\max\{C_i\}/\min\{C_i\}$. But these costs are dual costs, and we need to be careful they do not grow too large. The proof of lemma 4 uses the concavity of P_k. The proof of lemma 3 also depends crucially on properties of our dual solutions.

Recall that $c(e)$ is the capacity and $l(e)$ is the dual cost of edge e. Also recall that the dual cost $C_i(l)$ for flow i is given by $\sum_{e \in R_i} l(e)$ where R_i is the route of flow i. Let γ denote an upper bound on the cost of the dual solution obtained in the previous section.

For technical reasons, we will assume that the dual obtained in the previous section is multiplied by a scale factor s such that

$$\exists e \in E \ \sum_{j:e \in r_j} s\bar{x}_j = c(e)$$
$$\forall e \in E \ \sum_{j:e \in r_j} s\bar{x}_j \le c(e).$$

We will use the notation $x(l)$ for the final scaled dual solution, given edge costs l. The scale factor s can be obtained using a balanced tree; the details are similar to those in section 3 and are omitted.

Recall that $\beta_k(l)$ is the maximum value of $P_k(x)$ subject to the constraints $\{C(l)x = 1; x \ge 0\}$. After scaling, it is no longer necessarily true that $P_k(x(l)) \ge \beta_k(l)$ or that $C(l) \cdot x(l) \le \gamma$. However, it is still true that, for all k,

$$C(l) \cdot x(l) \le \frac{\gamma P_k(x(l))}{\beta_k(l)} \tag{1}$$

This is going to be sufficient for our purposes.

Let $D(l)$ denote the quantity $\sum_e l(e)c(e)$. We now state the algorithm for simultaneous primal optimization:

APPROX-FLOWS(ϵ, γ)
1   **for** each edge e
2   **do** $l_0(e) = \frac{\delta}{c(e)}$ // δ will be defined later.
3   $i = 0$
4   $x = 0$

5   **while** $D(l_i) < 1$
6   **do Flow phase:** Assume we are given a length function l_i.
7   Let $x(l_i)$ be the scaled γ-majorized solution for the dual problems.
8            $x = x + x(l_i)$
9   **Edge phase:**
10   **for** each edge e
11   **do** $f_i(e) = \sum_{j:e \in r_j} x_j(l_i)$
12                   $l_{i+1}(e) = l_i(e)(1 + \epsilon f_i(e)/c(e))$
13           $i = i + 1$
14   **return** x

We have already discussed how a scaled γ-majorized solution can be obtained in a distributed fashion. The stopping condition $D(l_i) < 1$ can also be checked easily using the same technique. Hence the above algorithm is amenable to an efficient distributed implementation.

We will now discuss the feasibility, the number of iterations, and the approximation ratio (i.e. the α) of the above algorithm. We will set $\epsilon = 1/2$ and $\delta = m^{-\frac{1}{\epsilon}}$. The proof for feasibility and time complexity mirrors that of Garg and Konemann; all the details which are particular to our problem manifest themselves in the analysis of the approximation ratio.

Feasibility of the Primal: Note that the solution returned by the above algorithm may be infeasible since there is always at least one saturated edge in any iteration. To make our flows feasible, we use the same trick as used by Garg and Konemann. Consider the increase of each edge's length. Since every new flow on edge e does not exceed its capacity, for every $c(e)$ units of flow routed through edge e, its length is increased by at least $1 + \epsilon$. Therefore for every edge e, $c(e)$ is overflowed by the factor of $\log_{1+\epsilon} \frac{l_{T-1}(e)}{l_0(e)}$ at most where T is the time when the algorithm terminates. Since $D(l_{T-1}) < 1$, $l_{T-1}(e) < 1/c(e)$. Therefore, flows divided by $\log_{1+\epsilon} \frac{1/c(e)}{\delta/c(e)} = \log_{1+\epsilon} \frac{1}{\delta}$ give us a feasible solution. In practice, we would multiply the flows obtained during each iteration by this feasibility factor, rather than doing it once at the end.

Running Time: Define the running time for computing the majorized dual solution as T_{sub}. Since our algorithm runs iteratively, and each iteration takes $O(T_{sub})$ time, the running time would be $O(T \cdot T_{sub})$ where T is the number of iterations to finish our algorithm.

In one iteration, we increase the length of a saturated edge by a factor of $(1 + \epsilon)$. Therefore, the number of iterations in which we increase the length of an edge e is at most $\left\lceil \log_{1+\epsilon} \frac{l_T(e)}{l_0(e)} \right\rceil$. Since the algorithm stops when $\sum_e l(e)c(e) > 1$, we have $l_T(e) \le 1/c(e)$, which gives

$$\log_{1+\epsilon} \frac{l_T(e)}{l_0(e)} \le \log_{1+\epsilon} \frac{1}{\delta}$$
$$= \tfrac{1}{\epsilon} \log_{1+\epsilon} m.$$

The above equation represents the number of iterations related to a particular edge. Multiplying by the total number of edges, m, yields

$$T \leq \left\lceil \frac{m}{\epsilon} \log_{1+\epsilon} m \right\rceil.$$

Since we have set ϵ to be $1/2$,

$$T = O(m \log m).$$

Approximation Guarantee: For brevity, let $D(i)$, $C(i)$, $\beta_k(i)$ and $x(i)$ denote $D(l_i)$, $C(l_i)$, $\beta_k(l_i)$ and $x(l_i)$ respectively.

Lemma 3. *Let P_k^* denote the optimum solution of the primal problem. When the algorithm terminates at time T,*

$$\frac{\sum_{j=1}^{T} P_k(x(j-1))}{P_k^*} \geq \frac{\ln \frac{1}{m\delta}}{\epsilon \gamma}.$$

Proof. For $i \geq 1$

$$
\begin{aligned}
D(i) &= \sum_e l_i(e) c(e) \\
&= \sum_e l_{i-1}(e) \left(1 + \epsilon \sum_{j:e \in r_j} x_j(i-1)/c(e)\right) c(e) \\
&= \sum_e l_{i-1}(e) c(e) + \epsilon \sum_e \left(l_{i-1}(e) \sum_{j:e \in r_j} x_j(i-1)\right) \\
&= D(i-1) + \epsilon \sum_j \left(\sum_{e \in r_j} l_{i-1}(e) x_j(i-1)\right) \\
&= D(i-1) + \epsilon \sum_j C_j(i-1) x_j(i-1) \\
&\leq D(i-1) + \frac{\epsilon \gamma P_k(x(i-1))}{\beta_k(i-1)} \qquad \text{(from equation 1)}
\end{aligned}
$$

Applying the duality theorem to the linear program for computing P_k^* yields

$$P_k^* \leq D(l) \beta_k(l).$$

The above step constitutes an important part of our proof. Notice that the duality theorem gives a different relation for each P_k. Informally. the guarantee obtained from equation 1 will exactly balances out this difference, resulting in the same guarantee for all dual problems. Now,

$$
\begin{aligned}
D(i) &\leq D(i-1) + \epsilon \gamma P_k(x(i-1)) \frac{D(i-1)}{P_k^*} \\
&= D(i-1) \left(1 + \epsilon \gamma \frac{P_k(x(i-1))}{P_k^*}\right).
\end{aligned}
$$

Hence,

$$
\begin{aligned}
D(i) &\leq \prod_{j=1}^{i} \left(1 + \epsilon \gamma \frac{P_k(x(j-1))}{P_k^*}\right) D(0) \\
&\leq \prod_{j=1}^{i} \left(e^{\epsilon \gamma \frac{P_k(x(j-1))}{P_k^*}}\right) m\delta \\
&= e^{\epsilon \gamma \sum_{j=1}^{i} \frac{P_k(x(j-1))}{P_k^*}} m\delta.
\end{aligned}
$$

Suppose our procedure stops at iteration T for which $D(T) \geq 1$. Then,

$$1 \leq D(T) \leq e^{\epsilon\gamma \sum_{j=1}^{T} \frac{P_k(x(j-1))}{P_k^*}} m\delta.$$

This implies

$$\frac{\sum_{j=1}^{T} P_k(x(j-1))}{P_k^*} \geq \frac{\ln \frac{1}{m\delta}}{\epsilon\gamma}.$$

We need to specify δ so that we get the approximation ratio to be appropriately bounded. The following theorem shows that our algorithm has a good bound on approximation ratio for an appropriate δ.

Lemma 4. *For* $\delta = m^{-\frac{1}{\epsilon}}$, *our algorithm generates at most* $\frac{\gamma}{(1-\epsilon)^2}$-*approximate solution.*

Proof. Define $S = \log_{1+\epsilon} 1/\delta$ which is the feasibility factor that every $x(i)$ should be divided by. In addition, define ρ to be $\frac{P_k^*}{P_k\left(\sum_{j=1}^{T} x(j-1)\right)/S}$ which is the approximation ratio. Since P_k is a concave function and $P_k(0) = 0$, we have

$$P_k\left(\sum_{j=1}^{T} x(j-1)\right) \geq \sum_{j=1}^{T} P_k(x(j-1)).$$

Therefore,

$$\begin{aligned}
\rho &\leq \frac{P_k^*}{\sum_{j=1}^{T} P_k(x(j-1))} \cdot S \\
&\leq \frac{\epsilon\gamma}{\ln \frac{1}{m\delta}} \cdot S && \text{(from lemma 3)} \\
&= \frac{\epsilon\gamma \log_{1+\epsilon} \frac{1}{\delta}}{\ln \frac{1}{m\delta}} \\
&= \gamma \frac{\frac{\epsilon}{\ln(1+\epsilon)}}{\ln \delta}.
\end{aligned}$$

Since $\ln(1 + \epsilon) \geq \epsilon(1 - \frac{\epsilon}{2})$ and $\frac{\ln \delta}{\ln(m\delta)} = \frac{1}{1-\epsilon}$ for $\delta = m^{-\frac{1}{\epsilon}}$,

$$\begin{aligned}
\rho &\leq \gamma \frac{1}{(1-\frac{\epsilon}{2})(1-\epsilon)} \\
&< \gamma \frac{1}{(1-\epsilon)^2}.
\end{aligned}$$

We set ϵ to be $1/2$. Then our desired approximation ratio would be 4γ. Recall that our bound on γ from theorem 4 is $O(\log n + \log \max\{C_i\}/\min\{C_i\})$. The quantities C_i are dual costs – we now relate them to primal capacities.

Lemma 5. *Define* R *to be* $\frac{\max_{e \in E} c(e)}{\min_{e \in E} c(e)}$. *Then,* $\log \max\{C_i\}/\min\{C_i\} = O(\log m + \log R)$.

Proof. Since the length of each edge is monotonically increasing with time, and C_i is just the sum of lengths,

$$\min_e l_0(e) \leq C_i \leq \sum_e l_{T-1}(e)$$

where T is the time when our algorithm terminates. Hence,

$$\max\{C_i\}/\min\{C_i\} \leq \frac{\sum_e l_{T-1}(e)}{\min_e l_0(e)}.$$

Since $l_0(e) = \delta/c(e)$, $\min_e l_0(e)$ is $\frac{\delta}{\max_e c(e)}$. On the other hand, note that our algorithm terminates when $D(l_T) > 1$, and hence $D(l_{T-1}) \leq 1$, which implies that $\sum_e l_{T-1}(e)c(e) \leq 1$. Thus, $\sum_e l_{T-1}(e) \leq \frac{1}{\min_e c(e)}$. Therefore,

$$\begin{aligned}
\max\{C_i\}/\min\{C_i\} &\leq \frac{\frac{1}{\min_e c(e)}}{\frac{\delta}{\max_e c(e)}} \\
&= \frac{\max_e c(e)}{\min_e c(e)} \cdot m^{1/\epsilon}.
\end{aligned}$$

Since we set $\epsilon = 1/2$,

$$\log \max\{C_i\}/\min\{C_i\} = O(\log m + \log R).$$

Our main result follows from the above sequence of lemmas, and is summarized by the following theorem:

Theorem 5. *The fractional packing framework of Garg and Konemann, combined with our distributed dual solution, results in an $O(\log n + \log R)$-majorized solution for the bandwidth allocation problem. The number of iterations required is $O(m \log m)$. Each iteration requires $O(\log n)$ message passing steps. Also, during each iteration, each flow-agent sends $O(1)$ messages of length $O(\log n + \log R)$ and performs at most $O(\log n + \log R)$ units of computation.*

It is important to note that the guarantee offered by the above theorem is the same (up to constant factors) as the existential guarantee offered by theorem 2 for this problem.

References

1. Y. Afek, Y. Mansour, and Z. Ostfeld. Convergence complexity of optimistic rate based flow control algorithms. *Proceedings of the 28th ACM Symposium on Theory of Computing*, pages 89–98, 1996.
2. B. Awerbuch and Y. Shavitt. Converging to approximated max-min flow fairness in logarithmic time. *Proceedings of the 17th IEEE Infocom conference*, pages 1350–57, 1998.
3. Y. Bartal, J. Byers, and D. Raz. Global optimization using local information with applications to flow control. *38th Annual Symposium on Foundations of Computer Science*, pages 303–312, 1997.
4. Y. Bartal, M. Farach-Colton, M. Andrews, and L. Zhang. Fast fair and frugal bandwidth allocation in atm networks. *Proceedings of the 10th Annual ACM-SIAM Symposium on Discrete Algorithms*, pages 92–101, 1999.
5. R. Bhargava, A. Goel, and A. Meyerson. Using approximate majorization to characterize protocol fairness. *Proceedings of ACM Sigmetrics*, pages 330–331, June 2001. (Poster paper).

6. N. Garg and J. Konemann. Faster and simpler algorithms for multicommodity flow and other fractional packing problems. *39th Annual Symposium on Foundations of Computer Science*, pages 300–309, 1998.

7. N. Garg and N. Young. On-line, end-to-end congestion control. *IEEE Foundations of Computer Science*, pages 303–312, 2002.

8. A. Goel and A. Meyerson. Simultaneous optimization via approximate majorization for concave profits or convex costs. *Technical report CMU-CS-02-203, Computer Science Department, Carnegie Mellon University*, December 2002.

9. A. Goel, A. Meyerson, and S. Plotkin. Approximate majorization and fair on-line load balancing. *Proceedings of the 12th ACM-SIAM Symposium on Discrete Algorithms*, pages 384–390, 2001.

10. A. Goel, A. Meyerson, and S. Plotkin. Combining fairness with throughput: On-line routing with multiple objectives. *Journal of Computer and Systems Sciences*, 63(1):62–79, 2001. A preliminary version appeared in ACM Symposium on Theory of Computing, 2000.

11. G.H. Hardy, J.E. Littlewood, and G. Polya. Some simple inequalities satisfied by convex functions. *Messenger Math.*, 58:145–152, 1929.

12. G.H. Hardy, J.E. Littlewood, and G. Polya. *Inequalities*. 1st ed., 2nd ed. Cambridge University Press, London and New York., 1934, 1952.

13. F.P. Kelly, A.K. Maulloo, and D.K.H. Tan. Rate control in communication networks: shadow prices, proportional fairness and stability. *Journal of the Operational Research Society*, 49:237–252, 1998.

14. J. Kleinberg, Y. Rabani, and E. Tardos. Fairness in routing and load balancing. *Proceedings of the 35th Annual Symposium on Foundations of Computer Science*, 1999.

15. A. Kumar and J. Kleinberg. Fairness measures for resource allocation. *Proceedings of 41st IEEE Symposium on Foundations of Computer Science*, 2000.

16. S. Low, L. Peterson, and L. Wang. Understanding TCP Vegas: a duality model. *Proceedings of ACM Sigmetrics*, 2001.

17. M. Luby and N. Nisan. A parallel approximation algorithm for positive linear programming. *Proceedings of 25th Annual Symposium on the Theory of Computing*, pages 448–57, 1993.

18. A.W. Marshall and I. Olkin. *Inequalities: theory of majorization and its applications*. Academic Press (Volume 143 of Mathematics in Science and Engineering), 1979.

19. S. Plotkin, D. Shmoys, and E. Tardos. Fast approximation algorithms for fractional packing and covering problems. *Math of Oper. Research*, pages 257–301, 1994.

20. A. Tamir. Least majorized elements and generalized polymatroids. *Mathematics of Operations Research*, 20(3):583–589, 1995.

Limits and Power Laws of Models for the Web Graph and Other Networked Information Spaces

Anthony Bonato[1] and Jeannette Janssen[2,*]

[1] Department of Mathematics, Wilfrid Laurier University,
Waterloo, ON, Canada, N2L 3C5
abonato@rogers.com
[2] Department of Mathematics and Statistics, Dalhousie University,
Halifax, NS, Canada, B3H 3J5
janssen@mathstat.dal.ca

Abstract. We consider a new model of the web graph and related networks. The model is motivated by the copying models of the web graph, where new nodes copy the link structure of existing nodes, and a certain number of additional random links are introduced. Our model parametrizes the number of random links, thereby allowing for the analysis of threshold behaviour. We consider infinite limits of graphs generated by our model, and compare properties of these limits with orientations of the infinite random graph. We analyze the power law behaviour of the in-degree distribution of graphs generated by our model.

1 Introduction to the Model

The overwhelming success of search engines that use a graph-based ranking system (Google being the most famous example) has made the *Web graph* a popular object of study. The Web graph is the graph formed by web pages or sites as nodes, and hyperlinks as directed edges. The Web graph is the most famous, and most complex, example of a *Networked Information Space (NIS)*. A NIS is a collection of information containing entities which are linked together. Other examples are digital libraries, which consist of publications linked by references [10], or networked databases recording phone calls or financial transactions.

In order to exploit the link structure of Networked Information Spaces, for ranking of search results, clustering, or focused crawling for example, we need to understand its generative process. Typically, the graphs in question have the property that their in-degree distributions satisfy a *power law*: for a positive integer k, the proportion of nodes of in-degree k is approximately $k^{-\gamma}$, where γ is an exponent that is generally observed to be between 2 and 3. This qualifies

---

* The authors gratefully acknowledges support from NSERC research grants, and from a MITACS grant. The first author acknowledges support from a Wilfrid Laurier University Senior Research Fellowship.

A. López-Ortiz and A. Hamel (Eds.): CAAN 2004, LNCS 3405, pp. 42–48, 2005.

them as so-called *scale-free networks*. Several stochastic models for the generation of scale-free networks have been proposed. In the *copying models* of [1, 11] a
new node copies part of the link environment of an existing node, while adding
an additional number of random links. Copy models seem to be especially appropriate to model the link structure of a NIS. Namely, it is very likely that a new
information entity (for example a web page or paper), is modelled on an existing
one, and hence, will create a link environment that will have a large overlap with
that of its model, but will also include some new links. Other recent models use
other paradigms (such as preferential attachment) for link creation; see [2, 3, 4].

We continue our study of [5] of the *infinite* limits of graphs generated by
stochastic models. One important reason why we consider such limits is to investigate the consequences of the choices made in the design of the model. In a
certain sense, the infinite limit magnifies the properties of the finite graphs that
lead to it. In our previous work, we considered generation processes for undirected graphs, and their measure of convergence to the *infinite random graph*,
or R (introduced by Erdős and Rényi in [9]). For a fixed positive integer, and
real number $p \in (0, 1)$, the graphs in $G(n, p)$ possess n nodes, and every pair of
distinct nodes is joined independently and with probability p. If we consider the
limit of the graphs in $G(n, p)$ as n tends to infinity, then the resulting graph will
with probability 1 be isomorphic to R. The graph R has a rich structure that
has been studied from a graph theoretic, algebraic, and topological perspective;
see the surveys [6, 7] for more information.

The graph R is undirected, as are the limiting graphs introduced in [5].
However, the link structure of a NIS often is by nature a directed graph: for
a web page, it is easy to see what pages it points to, but very hard to find
out what pages point to it. For scientific papers, it is definitely not the same
to cite or be cited. Hence, the question arises as to what infinite graphs could
be taken as directed versions of R. A suitable "generic" infinite directed graph
is the *Acyclic Random Oriented Graph*, or ARO, introduced and investigated
recently in [8]. ARO is an orientation of R. Like R, ARO is uniquely defined (up
to isomorphism) by a certain adjacency property (see Section 2), is the limit of
a sequence of finite acyclic directed graphs, and contains every acyclic directed
graph as an induced subgraph.

However, our research shows that graphs generated by the copy models are
unlikely to converge to ARO. As we prove in Theorem 3, the limits are *isomorphic* to ARO with low probability, but in some cases are *homomorphically
equivalent* with ARO with high probability. In order to classify the limits that we
do obtain, we introduce the concept of a *near*-ARO. There are infinitely many
non-isomorphic near-ARO's; this could imply a certain sensitivity of the copying
model to initial conditions. Near-ARO's have a rich structure; see Theorem 1.
While convergence to ARO may be viewed as a process of loosing all structure,
convergence to a near-ARO may be viewed as loosing a *significant amount* of
structure.

We introduce a model $D(p, \rho, H)$ motivated by the copying models. Like the
copying models, our model uses copying of the existing link structure as its

basic rule for link creation. Our model allows for variation of the number of random links. The three parameters of this model are a fixed *copying probability* $p \in (0, 1)$, a *random link function* $\rho : \mathbb{N} \to \mathbb{R}$, defined by

$$\rho(t) = \alpha t^s,$$

where α and s are non-negative real constants so that $s \in [0, 1]$, and a fixed finite acyclic *initial digraph* H.

1. At $t = 0$, set $G_0 = H$.
2. For a fixed $t \geq 0$, assume that G_t has been defined, is finite, and contains G_0 as an induced subdigraph. To form G_{t+1}, add a new node v_{t+1} to G_t and choose its out-neighbours as follows.
   (a) Choose an existing node u from G_t uniformly at random (u.a.r.). The node u is called the *copying node*.
   (b) For each out-neighbour w of u, independently add a directed edge from v_{t+1} to w with probability p. In addition, choose $\lfloor \rho(t) \rfloor$-many (not necessarily distinct) nodes from $V(G_t)$ u.a.r., and add directed edges from v_{t+1} to each of these nodes. The latter edges are called *random links*.
   (c) Make the digraph G_{t+1} simple by removing any parallel edges.

At each time-step t, our model adds approximately $\rho(t)$-many random links between the new node and the existing nodes. Theorem 3 shows that if $\rho(t) \in \theta(t)$, then the limit is a near-ARO with high probability. As s tends to 1, with high probability the limit, while not a near-ARO, behaves increasingly like a near-ARO. On the other hand, Theorem 5 shows that we loose power law behaviour if we have more than a constant number of random links. Hence, an interesting "grey area" where $0 < s < 1$ emerges; we will elaborate further on this in Section 4.

2   Limits and the $D(p, \rho, H)$ Models

Before we state the main results for this section, we require a few definitions. If u is a node in a digraph G, then let

$$N_\uparrow(u) = \{v \in V(G) : (u, v) \in E(G)\}$$

be the *out-neighbourhood* of u in G. If $(G_t : t \in \omega)$ is a sequence of digraphs with G_t an induced subdigraph of G_{t+1}, then define the *limit* of the G_t, written

$$G = \lim_{t \to \infty} G_t,$$

by

$$V(G) = \bigcup_{t \in \mathbb{N}} V(G_t),$$

and

$$E(G) = \bigcup_{t \in \mathbb{N}} E(G_t).$$

A digraph G is *good* if G is acyclic, has no infinite directed paths, and each node of G has finite out-degree. For example, if G_t is generated by our model $D(p, \rho, H)$, then any limit $G = \lim_{t \to \infty} G_t$ is good. A digraph G is an *acyclic random oriented graph*, or *ARO* for short, if G is good, and for each finite set $S \subset V(G)$, there are infinitely many nodes u such that $S = N_\uparrow(u)$. AROs were introduced and investigated recently in [8], where it was proved (among other things) that a countable ARO is unique up to isomorphism. Hence, we will refer to this unique isomorphism-type simply as ARO. (Strictly speaking, we are using the *inverse* of ARO as defined in [8], where the inverse of a digraph results by reversing the orientations of all the directed edges. Since we have only have use for ARO as defined above, we will keep our notation.) As noted first in [8], ARO results from a suitable orientation of the infinite random graph R. Indeed, ARO may be defined probabilistically: let $\mathbb{N}$ be the set of nodes, allow all edges to be directed backward (that is, (i, j) is a directed edge only if $j < i$), and adopt these edges independently with probability 2^{i+j}. The digraph ARO results with probability 1 from this random digraph.

We say that a digraph G is a *near-ARO* if it is good, and for each finite set $S \subset V(G)$, there is a node $u \in V(G)$ such that $S \subseteq N_\uparrow(u)$. ARO is clearly near-ARO. However, there are many examples of countable near-ARO digraphs that are not isomorphic to ARO; see Corollary 1 below.

We say that an undirected graph G is *algebraically closed*, or *a.c.* for short, if for each finite subset U of nodes of G, there is a node $z \in V(G) \setminus U$ such that z is joined to each node of U. For example, an infinite clique and R are examples of a.c. graphs. A *homomorphism* from the digraph G to H is an edge-preserving mapping from $V(G)$ to $V(H)$. The digraphs G and H are *homomorphically equivalent*, written $G \leftrightarrow H$, if there is a homomorphism from G to H and from H to G. Note that isomorphic digraphs are homomorphically equivalent, although the converse fails.

The following theorem (whose proof relies on König's infinity lemma and the back-and-forth method, and so is omitted) characterizes near-ARO digraphs up to homomorphic equivalence.

Theorem 1. *Let $G = \lim_{t \to \infty} G_t$ be a good digraph. The following are equivalent.*

1. *The underlying graph of G (formed by forgetting the orientation of each directed edge) is a.c.*
2. *The digraph G is a near-ARO.*
3. *The digraph $G \leftrightarrow ARO$.*
4. *For all countable good digraphs H, H admits a homomorphism into G.*

While the digraph ARO is unique up to isomorphism, the following corollary demonstrates that the maximum possible number of non-isomorphic near-ARO digraphs exist. We write $2^{\aleph_0}$ for the cardinality of the set of real numbers.

Corollary 1. *There are $2^{\aleph_0}$ many non-isomorphic countable near-ARO digraphs.*

We say that a digraph satisfies the *locally near-ARO* adjacency property if it is good, and for all finite sets of nodes S that are in the out-neighbourhood of some other node y, there is a node whose out-neighbours include S. Clearly, a near-ARO digraph is locally near-ARO; Theorem 3 (4) will demonstrate that the converse is false. Our next result shows that for all values of s, limits of graphs generated by $D(p, \rho, H)$ are locally near-ARO with high probability.

Theorem 2. *Fix $p \in (0, 1)$, ρ, and H. With probability 1, the limit*

$$G = \lim_{t \to \infty} G_t$$

of graphs generated by the model $D(p, \rho, H)$ is locally near-ARO.

Proof. Since a countable union of measure 0 subsets has measure 0, it suffices to show that for a fixed $y \in V(G)$ and a finite $S \subseteq N_\uparrow(y)$ the probability that there is no node joined to all of S is 0 (since there only countably many choices for y and S in G).

Let t_0 be the least integer such that y and S are in $V(G_{t_0})$. Let $|V(G_{t_0})| = m$ and $|S| = i$. If $t \geq t_0$, the probability that y is chosen as copying node in G_t equals $\frac{1}{m+t-t_0}$. Given that y is the copying node, v_t is joined to all of S with probability p^i. Then the probability that no node of G is joined to all of S is at most

$$\prod_{t=t_0}^{\infty} \left(1 - \left(\frac{1}{m+t-t_0}\right) p^i\right) = 0. \qquad \square$$

Our main result is the following theorem, which demonstrates that as s tends to 1, graphs G generated by $D(p, \rho, H)$ share more and more properties of a near-ARO. Further, the graphs G are very rarely isomorphic to ARO. For a positive integer n, we say that a digraph G is *n-near-ARO* if it is good, and for each set $S \subset V(G)$ of cardinality n, there is a node $u \in V(G)$ such that $S \subseteq N_\uparrow(u)$. Observe that a digraph is near-ARO if and only if it is n-near-ARO for all positive integers n.

Theorem 3. *Fix $p \in (0, 1)$, $\rho = \alpha t^s$, and H. Let G be the limit of a sequence of digraphs generated according to the model $D(p, \rho, H)$.*

1. *If $s = 1$, then with probability 1 G is near-ARO.*
2. *If $s \in [0, 1)$, then with probability 1 G is $\lfloor \frac{1}{1-s} \rfloor$-near-ARO.*
3. *If $s \in [0, 1)$, then with positive probability G is not near-ARO.*
4. *For all $s \in [0, 1]$, with probability 1 G is not isomorphic to ARO.*

Theorem 3 suggests a *threshold* behaviour for convergence to a near-ARO: as s tends to 1, with high probability the limit G acquires more and more properties of a near-ARO, but with positive probability is not near-ARO. At $s = 1$, we obtain a near-ARO with high probability.

Proof of Theorem 3. We sketch a proof of (2) only. Let $G = \lim_{t \to \infty} G_t$. It is straightforward to see that G is good. As in the proof of Theorem 2, it suffices to show that for a fixed finite $S \subseteq V(G)$ the probability that there is no node joined to all of S is 0.

Fix S a set of nodes with $|S| \leq \left\lfloor \frac{1}{1-s} \right\rfloor$, and let t_0 be the first t so that each node of S belongs to $V(G_t)$. We can prove that for all $t \geq t_0$, the probability that v_t is joined to every node of S is at least

$$\beta(t-1)^{(s-1)|S|}(1+o(1)) \geq \beta(t-1)^{-1}(1+o(1)), \tag{1}$$

where $\beta \in (0,1)$ is a constant that does not depend on t. Hence, by (1), the probability that there is no node of G joined to every node of S is at most

$$\prod_{t=t_0}^{\infty} \left(1 - \beta(t-1)^{-1}(1+o(1))\right) = 0. \qquad \square$$

Theorem 3 (1) may be generalized to other values of $\rho(t)$ (which are not necessarily a power of t) with only minor changes in the proof.

Theorem 4. *Let G be the limit of a sequence of graphs generated according to the model $D(p, \rho, H)$, with ρ a non-negative, monotone increasing function $\rho : \mathbb{N} \to \mathbb{R}$ satisfying the condition:*

$$\sum_{t=0}^{\infty} \frac{\rho(t)}{t^2} = \infty. \tag{2}$$

Then with probability 1, G is near-ARO.

3   Degree Distributions of the $D(p, \rho, H)$ Models

When do the models $D(p, \rho, H)$ produce digraphs whose in-degree distributions follow power laws? We find in the following results that power laws are sensitive to the choice of ρ.

Theorem 5. *Fix $p \in (0, 1)$ and H. Let G be the limit of a sequence of graphs generated according to the model $D(p, \rho, H)$, where $\rho(t) = \alpha t^s$. Then the degree distribution of G_t converges to a power law distribution if and only if $s = 0$.*

Proof. Let $X_i(t)$ be the expected number of nodes of in-degree i at time t. Suppose that

$$\lim_{t \to \infty} \frac{X_i(t)}{t} = b_i = ci^{-d},$$

for some positive constants c and d. It follows that

$$b_i = p((i-1)b_{i-1} - ib_i) + \lfloor \rho(t) \rfloor (b_{i-1} - b_i) + o(1).$$

By definition b_i is a constant, so either $b_i - b_{i-1} = 0$ or $\rho(t) = \alpha t^0 = \alpha$. If $b_i = b_{i-1}$, then this contradicts that $b_i = ci^{-d}$. Therefore, $s = 0$ and $\rho(t) = \alpha$. We omit the details that if $\rho(t) = \alpha$, then a power law is obtained. $\qquad \square$

4   Conclusions and Future Work

We introduced a new model $D(p, \rho, H)$ of the web graph and other Networked Information Spaces, motivated by the copying models of the web graph proposed by [1, 11]. $D(p, \rho, H)$ provides a continuum of models, whose structural properties depend largely on the number of random links parameterized by $\rho = \alpha t^s$.

We have seen that for all values of $s \in [0, 1]$, our model generates limit graphs G which satisfy the locally-near-ARO adjacency property. As s tends to 1, G becomes increasingly random, until at $s = 1$ it is with probability 1 homomorphically equivalent with a certain random acyclic digraph, ARO. Hence, on the one hand, the model $D(p, \rho, H)$ is robust: a large number of random links must be added at each time-step to ensure a random-like structure. Further, the choices of p and H seem to have little impact on the structure of the limit. On the other hand, we obtain power laws only if there are at most a constant number of random links. Hence, for any choice of ρ with $0 < s < 1$, there is an interesting "grey area" that emerges: the limits are not completely random, nor do we obtain power laws. We do not understand exactly the in-degree distributions that arise when $s \in (0, 1)$. We plan on analyzing these distributions in future work.

References

1. M. Adler, M. Mitzenmacher, Towards compressing web graphs, In: *Poceedings of the IEEE Data Compression Conference (DCC)*, 2001.
2. W. Aiello, F. Chung, L. Lu, Random evolution in massive graphs, Handbook on Massive Data Sets, (Eds. James Abello et al.), Kluwer Academic Publishers, (2002) 97-122.
3. R. Albert, A. Barabási, Emergence of scaling in random networks, *Science* **286** (1999) 509-512.
4. B. Bollobás, O. Riordan, J. Spencer, and G. Tusnády, The degree sequence of a scale-free random graph process, *Random Structures Algorithms* **18** (2001) 279-290.
5. A. Bonato, J. Janssen, Infinite limits of copying models of the web graph, *Internet Mathematics* **1** (2003) 193-213.
6. P.J. Cameron, The random graph, in: *Algorithms and Combinatorics* **14** (R.L. Graham and J. Nešetřil, eds.), Springer Verlag, New York (1997) 333-351.
7. P.J. Cameron, *The random graph revisited*, in: European Congress of Mathematics, Vol. I (Barcelona, 2000), 267–274, *Progr. Math.*, 201, Birkhuser, Basel, 2001.
8. R. Diestel, I. Leader, A. Scott, and S. Thomassé, Partitions and orientations of the Rado graph, submitted.
9. P. Erdős and A. Rényi, Asymmetric graphs, *Acta Math. Acad. Sci. Hungar.* **14** (1963) 295-315.
10. Y. An, J. Janssen, E. Milios, Characterizing the citation graph of computer science literature, accepted in *Knowledge and Inf. Systems (KAIS)*.
11. R. Kumar, P. Raghavan, S. Rajagopalan, D. Sivakumar, A. Tomkins, and E. Upfal, Stochastic models for the web graph, In: *Proceedings of the 41th IEEE Symp. on Foundations of Computer Science*, 2000.

Cuts and Disjoint Paths in the Valley-Free Path Model of Internet BGP Routing[*]

Thomas Erlebach[1], Alexander Hall[2,**], and Alessandro Panconesi[3], and Danica Vukadinović[2,**]

[1] Dept. of Computer Science, University of Leicester, Leicester LE1 7RH, UK
t.erlebach@mcs.le.ac.uk
[2] Computer Engineering and Networks Laboratory (TIK),
Department of Information Technology and Electrical Engineering,
ETH Zurich, Switzerland
{hall, vukadin}@tik.ee.ethz.ch
[3] DSI – Università La Sapienza, Rome, Italy
ale@dsi.uniroma1.it

Abstract. In the valley-free path model, a path in a given directed graph is valid if it consists of a sequence of forward edges followed by a sequence of backward edges. This model is motivated by BGP routing policies of autonomous systems in the Internet. Robustness considerations lead to the problem of computing a maximum number of disjoint paths between two nodes, and the minimum size of a cut that separates them. We study these problems in the valley-free path model. For the problem of computing a maximum number of edge- or vertex-disjoint valid paths between two given vertices s and t, we give a 2-approximation algorithm and show that no better approximation ratio is possible unless $P = NP$. For the problem of computing a minimum vertex cut that separates s and t with respect to all valid paths, we give a 2-approximation algorithm and prove that the problem is APX-hard. The corresponding problem for edge cuts is shown to be polynomial-time solvable. We present additional results for acyclic graphs.

1 Introduction

Let $G = (V, E)$ be a directed graph. For $s, t \in V$, a path from s to t is *valid* if it consists of a (possibly empty) sequence of forward edges followed by a (possibly empty) sequence of backward edges. Note that a valid path from s to t gives also a valid path from t to s and vice versa. We refer to this model of valid paths as the *valley-free path model*. The reason for this terminology is that if we view

---

[*] Research partially supported by the European Commission in the 5th Framework Programme under contract IST-2001-32007 (APPOL II) and in the 6th Framework Programme under contract 001907 (DELIS), with funding for the Swiss participants provided by the Swiss Federal Office for Education and Science.
[**] Supported in DICS-Project No. 1838 by the Hasler Foundation.

A. López-Ortiz and A. Hamel (Eds.): CAAN 2004, LNCS 3405, pp. 49–62, 2005.
© Springer-Verlag Berlin Heidelberg 2005

directed edges as "pointing upward" towards their heads, a path is valid if and only if it does not contain a "downward" edge followed by an "upward" edge, i.e., a valley ($\bigvee$).

The motivation for studying the valley-free path model comes from BGP routing policies in the Internet on the level of autonomous systems, as explained in Section 1.1 in more detail. Robustness considerations of the Internet topology then lead naturally to questions concerning the computation of large sets of disjoint valid paths between two given vertices, and of small vertex or edge cuts separating two given vertices with respect to all valid paths. The corresponding optimization problems for standard directed paths can be solved efficiently using network flow techniques (e.g., see [1]). In this paper, we initiate the investigation of these problems in the valley-free path model. It turns out that several of these problems are *NP*-hard in this model. Our main results are:

- We give 2-approximation algorithms for the problems of computing a maximum number of vertex- or edge-disjoint valid paths between two given vertices s and t, and we show that it is *NP*-hard to approximate these problems within ratio $2 - \varepsilon$ for any fixed $\varepsilon > 0$.
- We prove *APX*-hardness for the problem of computing a min valid s-t-vertex-cut, i.e., a minimum size set of vertices whose removal from G disconnects all valid paths between s and t. We also give a 2-approximation algorithm for this problem.
- For the edge version of the problem, i.e., computing a min valid s-t-edge-cut, we give a polynomial algorithm that computes an optimal solution.
- We prove that the size of a min valid s-t-cut is at most twice the maximum number of disjoint valid s-t-paths, both for the edge version and the vertex version of the problems, and we show that this bound is tight.
- For the special case that the given graph G is acyclic (where "acyclic" is to be understood in the standard sense, i.e., the directed graph G is acyclic if it does not contain a directed cycle), we obtain a polynomial algorithm for finding k edge- or vertex-disjoint valid paths between s and t if they exist, where k is an arbitrary constant. The algorithm is based on a generalization of ideas due to Fortune, Hopcroft and Wyllie [9]. We also establish *NP*-hardness for the general problem of computing a maximum number of vertex- or edge-disjoint valid s-t-paths in acyclic graphs.

Our results provide interesting insights into natural variations of the classical problems of computing disjoint s-t-paths and minimum s-t-cuts. Furthermore, the algorithms we provide may be useful for investigating issues related to the robustness of the Internet topology while taking into account the effects of routing policies.

1.1   Motivation: Autonomous Systems on the Internet

In this section we discuss the issues in Internet routing on the autonomous system level that have motivated our study. An autonomous system (AS) on the Internet is a subnetwork under separate administrative control. ASs are connected by physical links and exchange routing information using the Border Gateway Protocol (BGP). An AS can consist of tens to thousands of routers and hosts. On the level of ASs, the Internet can be represented as an undirected graph with a vertex for each AS and an edge between two ASs if they have at least one physical link between them. But such an undirected graph is not sufficient to model the effects of routing policies enforced by individual ASs. Each AS announces the routes to a certain set of destination ASs (more precisely, address prefixes) to some of its neighbors. The decisions which routes to announce to which neighbors are determined by BGP routing policies. These depend mostly on the economic relationships between ASs and represent an important aspect of the Internet structure.

The nature of commercial agreements between ASs has attracted a lot of attention in the Internet economics research community [12, 13, 3]. The main trends in the diversity of these agreements were described in [12, 13]. The impact of economic relationships on the engineering level, more precisely on BGP routing, has not been immediately recognized despite the direct implication that an existing link between two ASs will not be used to transfer traffic that collides with their mutual agreement. Then several papers showing the impact of BGP policies on features such as path inflation and routing convergence have appeared [16, 17, 14].

Consequently, the previously developed undirected model for the AS topology is not satisfactory because it allows some prohibited paths between ASs and thus might produce a distorted picture of BGP routing. On the other hand, involving all of the peculiarities of the contracts between autonomous systems in a new model would add too much complexity. Thus, a rough classification into a few categories was proposed for the BGP policies adopted by a pair of ASs: customer-provider, peer-to-peer, and siblings (see [10]). Later on, a simplified model with only two categories (customer-provider and peer-to-peer) was proposed [15].

A customer-provider relationship between A and B can be represented as a directed edge from A to B, and a peer-to-peer relationship as an undirected edge. If ASs A and B are in a customer-provider relationship, B announces all its routes to A, but A announces to B only its own routes and routes of its own customers. If they are peers, they exchange their own routes and routes of their customers, but not routes that they learn from their providers or other peers. This leads to the model proposed in [15] that a path is valid if and only if it consists of a sequence of customer-provider edges ($\bullet\rightarrow\bullet$), followed by at most one peer-to-peer edge ($\bullet-\bullet$), followed by a sequence of provider-customer edges ($\bullet\leftarrow\bullet$).

It is easy to see that a peer-to-peer edge (undirected edge) between A and B can be replaced by two customer-provider edges from A to X and from B to X, where X is a new node, without affecting the solutions to any of the optimization problems (minimum cut problems and maximum disjoint paths problems) we study in this paper. Therefore, without losing generality, we can consider a model with only customer-provider edges. In other words, this model consists of a directed graph with ASs as nodes and where the edge-directions represent economic relationships. Here the valley-free paths are exactly the paths permitted by the BGP routing policies.

Information about the economic relationships between autonomous systems is not publicly available. Therefore, several approaches to inferring these relationships from available topology data or AS path information have been proposed in the literature [10, 15, 8, 6].

If a communication network is represented as an undirected or directed graph in a model without routing policies, it is natural to measure the connectivity provided to an s-t-pair as the maximum number of disjoint s-t-paths or the minimum size of an s-t-cut (by Menger's theorem, these two quantities are the same). This motivates us to study the corresponding notions for the valley-free path model in this paper.

It seems natural to expect that the directed graph of customer-provider edges is acyclic (i.e., does not contain a directed cycle), because providers should be "higher" in the Internet hierarchy than their customers. But it turns out that the graphs obtained with several of the abovementioned algorithms do contain directed cycles. Thus, we are interested in cuts and disjoint paths both in general directed graphs and in acyclic graphs.

1.2   Outline

In Section 2, we give the necessary definitions and discuss some preliminaries. Section 3 contains our results for disjoint paths and minimum cuts in general directed graphs. In Section 4, we consider acyclic graphs. We give our conclusions and point to some open problems in Section 5. Proofs omitted due to space restrictions can be found in [7].

2   Preliminaries

Following the terminology from [15, 6, 8], where the problem of classifying the relationships between ASs is called the Type-of-Relationship (ToR) problem, we call a simple directed graph $G = (V, E)$ a *ToR graph* if G has no loops and no anti-parallel edges (i.e., $(u, v) \in E$ implies $(v, u) \notin E$). In terms of the underlying motivation, a directed edge from u to v, where $u, v \in V$, means that u is a customer of v. A path $p = v_1, v_2, \cdots, v_r$ in a ToR graph is *valid* (and called a valid v_1-v_r-path), if it satisfies the following condition:

> There exists some j, $1 \leq j \leq r$, such that $(v_i, v_{i+1}) \in E$ for $1 \leq i \leq j-1$ and $(v_i, v_{i-1}) \in E$ for $j + 1 \leq i \leq r$.

The part of the path from v_1 to v_j is called its *forward part*, the part from v_j to v_r its *backward part*. Note that valid paths are symmetric, i.e., the reverse of a valid s-t-path is a valid t-s-path. The existence of a valid s-t-path can be checked in linear time by performing a standard directed depth-first search from s and from t and testing if any vertex is reachable from both s and t along a directed path.

Let $s, t \in V$ be two distinct vertices in a ToR graph $G = (V, E)$. A set $C \subseteq V \setminus \{s, t\}$ is a *valid s-t-vertex-cut* if there is no valid path from s to t in $G - C$. A smallest such set C is called a *min valid s-t-vertex-cut*. The *min valid s-t-edge-cut* is defined analogously. Two valid s-t-paths are called vertex-disjoint (edge-disjoint) if the only vertices that they have in common are s and t (if they have no edges in common). The optimization problems that we are interested in are those of computing minimum size cuts and maximum size sets of disjoint paths, both in the vertex version and in the edge version: the min valid s-t-vertex-cut problem, the min valid s-t-edge-cut problem, the max vertex-disjoint valid s-t-paths problem, and the max edge-disjoint valid s-t-paths problem. An approximation algorithm A for an optimization problem Π is a polynomial algorithm that always outputs a feasible solution. A is a ρ-approximation algorithm (has approximation ratio ρ), if for all inputs I, $OPT(I)/A(I) \leq \rho$, if Π is a maximization problem, or $A(I)/OPT(I) \leq \rho$, if Π is a minimization problem. Here, $OPT(I)$ is the objective value of an optimal solution, and $A(I)$ is the objective value of the solution computed by A, for a given input I. APX is the class of all optimization problems (with some natural restrictions [2]) that can be approximated within a constant factor. A problem is APX-hard if every problem in APX can be reduced to it via an approximation preserving reduction. For every APX-hard problem there is a constant $\rho > 1$ such that it is not possible to find a ρ-approximation algorithm for the problem unless $P = NP$. See [2] for further information about approximability classes and approximation preserving reductions.

3   Results for General ToR Graphs

First, we introduce a helpful *two-layer model* that leads to a relaxation of flows and cuts in ToR graphs.

3.1   The Two-Layer Model

From a ToR graph $G = (V, E)$ and $s, t \in V$ we construct a *two-layer model H*, which is a directed graph, in the following way. Two copies of the graph G are made, called the *lower* and the *upper layer*. In the upper layer all edge-directions are reversed. Every node v in the lower layer is connected with an edge to the corresponding copy of v, denoted v', in the upper layer. The edge is directed from v to v'. (When dealing with edge-cuts or edge-disjoint paths, we let H contain $|V|$ parallel copies of the edge (v, v') to ensure that these "vertical" edges are not contained in minimum edge-cuts and do not restrict the number of edge-disjoint paths switching from the lower to the upper layer at v.) Finally, we obtain the two-layer model H by identifying the two s-nodes (of lower and upper layer) and also the two t-nodes, and by removing the incoming edges of s and the outgoing edges of t.

A valid path $p = v_1, \cdots, v_r$ in G with $v_1 = s$ and $v_r = t$ is equivalent to a directed path in H in the following way. The forward part of p, i.e., all edges $(v_i, v_{i+1}) \in p$ that are directed from v_i to v_{i+1}, is routed in the lower layer. Then

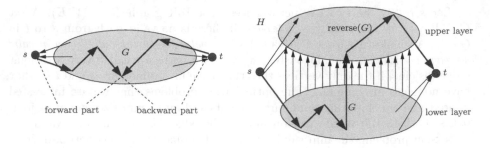

Fig. 1. Path in ToR graph G and corresponding path in the two-layer model H

there is a possible switch to the upper layer with a (v, v') type edge (there can be at most one such switch). The backward part of p is routed in the upper layer. If there is only a forward respectively a backward part of p, then the corresponding path in H is only in the lower respectively upper layer. See Fig. 1 for an example.

We now detail in which sense the two-layer model yields relaxations of vertex-respectively edge-disjoint paths and vertex- respectively edge-cuts in ToR graphs.

Note that two vertex-disjoint valid paths in G directly give two vertex-disjoint paths in H. But two vertex-disjoint paths p_1, p_2 in H do not necessarily correspond to vertex-disjoint valid paths in G. The path p_1 might use the node v and the path p_2 its counterpart v' in the other layer, yielding two valid paths that are not vertex-disjoint in G. The analogous statements apply to edge-disjoint paths.

A valid s-t-vertex-cut in G directly gives an s-t-vertex-cut in H of twice the cardinality: simply take for each cut node in G the corresponding nodes from both layers in H. But, of course, there might be an s-t-vertex-cut in H without the property that for each node v in the cut, also its counterpart v' is in the cut. Analogous statements apply to edge-cuts.

3.2   Min Valid s-t-Vertex-Cut

First, we are able to establish the hardness of the min valid s-t-vertex-cut problem by a reduction from the 3-way edge cut problem in undirected graphs.

Theorem 1. *For a given ToR graph $G = (V, E)$ and $s, t \in V$, finding the min valid s-t-vertex-cut is NP-hard and even APX-hard.*

Proof. We use a similar technique as in [11], reducing the undirected 3-way edge cut problem to the min valid s-t-vertex-cut problem in ToR graphs. In the undirected 3-way edge cut problem, we are given an undirected graph G, and three terminals v_1, v_2, v_3. The goal is to find a minimum set of edges in G such that after removing this set, all pairs of vertices in $\{v_1, v_2, v_3\}$ are disconnected. This problem is proven to be NP-hard in [5].

Let $G = (V, E)$ be such an undirected graph with 3 distinct terminals v_1, v_2 and v_3. We create a ToR graph G' in the following way: each node v of G is replaced with $\deg(v)$ copies of the same node. For each edge $\{u, w\}$ in G, a gadget consisting of 2 new nodes, $e_1^{u,w}$ and $e_2^{u,w}$, is added. The gadget includes an edge from $e_1^{u,w}$ to $e_2^{u,w}$, edges from all copies of u and w to $e_1^{u,w}$ and from $e_2^{u,w}$ to all copies of u and w. We also add two nodes s and t and the edges from s to all copies of v_1, from all copies of v_2 to s and t, and from t to all copies of v_3. See Fig. 2 for a simple example.

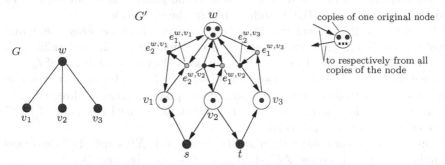

Fig. 2. Example of the transformation of the original undirected graph G to the ToR graph G'

Note that every valid path between any copy of u and any copy of w via the gadget added for the edge $\{u, w\}$ contains $e_1^{u,w}$. This holds because no path "copy of w, $e_2^{u,w}$, copy of u" or "copy of u, $e_2^{u,w}$, copy of w" is valid.

In the following we will first show that any valid s-t-vertex-cut in G' can be transformed into a cut of at most the same cardinality that only contains e_1 type nodes. Then we prove that there is a direct correspondence between 3-way edge cuts in G and valid s-t-vertex-cuts in G' that contain only such e_1 type nodes. This yields in particular that an approximation algorithm for the min valid s-t-vertex-cut problem gives an approximation algorithm of the same ratio for the 3-way edge cut problem.

Assume we are given a valid s-t-vertex-cut C in G' that contains nodes that are not of type e_1. We can assume that C is a minimal cut (inclusion-wise). If the cut contains a copy u' of u, where u is a node in the original graph G, it must also contain all other copies of u. Otherwise there is always an equivalent "detour" path via one of these other copies, rendering the addition of u' to C superfluous and contradicting the minimality of C. If C contains all $\deg(u)$ copies of a node u, we replace these nodes in C by the e_1 type nodes of the neighboring gadgets. There are exactly $\deg(u)$ such nodes. Every valid s-t path containing a copy of u traverses at least one of these neighboring gadgets (and therefore its e_1 node, see above). To see this, note that the paths "s, copy of v_2, t" and "t, copy of v_2, s" are not valid. Thus by this replacement we did not reintroduce previously cut paths. The cardinality of C did not increase. If C contains an e_2 type node, we replace it by the corresponding e_1 type node. Once more the cardinality of C does not increase and no valid paths are reintroduced. Now C contains only e_1 type nodes.

Next, we show that for a set of edges Q in the graph G, the corresponding set of nodes $C = \{e_1^{u,w} | \{u,w\} \in Q\}$ in G' is a valid s-t-vertex-cut if and only if Q is a 3-way cut for terminals v_1, v_2 and v_3 in G. (As previously mentioned, a 3-way cut disconnects all possible pairs in $\{v_1, v_2, v_3\}$.)

Note that the gadgets ensure that for any two nodes u and w in G' corresponding to the endpoints of an undirected edge $\{u, w\}$ in G, there exists a directed path from u to w and from w to u in the gadget added for $\{u, w\}$. To disconnect all such u-w paths, it suffices to cut the $e_1^{u,w}$ node.

First, if C is a valid s-t-vertex-cut, Q is a 3-way cut, because otherwise, if any pair of terminals v_1, v_2, v_3 are connected in G, then there will be a valid path between s and t which will give a contradiction. On the other hand, if Q is a 3-way cut, there is no valid path between s and t in $G' - C$, because at least one gadget is disconnected on every valid path corresponding to an undirected path between v_i and v_j for $i \neq j$ in G and, as noted before, the paths "s, copy of v_2, t" and "t, copy of v_2, s" are not valid.

Thus, we have shown that min valid s-t-vertex-cut is NP-hard. APX-hardness also follows directly from the APX-hardness of 3-way edge cut [5]. □

Note that the proof does not carry over to min valid s-t-edge-cut, because no gadget can be found where the role of the $e_1^{u,w}$ node is taken over by exactly one edge (such that the copies of u and the copies of w are disconnected if this edge is deleted). In fact, in Section 3.3 we will show that a polynomial-time optimal algorithm exists for the min valid s-t-edge-cut problem.

A Simple 2-Approximation. Given a ToR graph $G = (V, E)$ and $s, t \in V$ (where we assume that there is no direct edge in G between s and t, because otherwise a valid s-t-vertex-cut does not exist), the min valid s-t-vertex-cut approximation algorithm is as follows:

ALGORITHM VERTEXCUT

---

1. From G construct the two-layer model H as described in Section 3.1.
2. Compute a min s-t-vertex-cut C_H in H.
3. Output the set $C_G = \{v \in V |$ at least one copy of v is in $C_H\}$ as valid s-t-vertex-cut.

Clearly $|C_G| \leq |C_H|$ holds and C_G is a valid s-t-vertex-cut in G. Let C_{opt} be a min valid s-t-vertex-cut in G. As mentioned in Section 3.1, by duplicating C_{opt} for both layers of H one obtains an s-t-vertex-cut in H. Thus $|C_H|$ is at most twice $|C_{opt}|$, which gives Theorem 2.

Theorem 2. *There is a 2-approximation algorithm for the min valid s-t-vertex-cut problem in ToR graphs.*

3.3   Min Valid s-t-Edge-Cut

Quite surprisingly, there is a polynomial-time optimal algorithm for the min valid s-t-edge-cut problem in ToR graphs. This is in contrast to many flow and cut problems in directed and undirected graphs, where the node and the edge variant of the respective problem are of the same complexity.

Let EDGECUT be the reformulation of algorithm VERTEXCUT from Section 3.2 that considers edges instead of vertices. (Recall that for the edge version of the cut problem, we also modify the 2-layer model H in such a way that each vertex in the lower layer is connected to its copy in the upper layer by $|V|$ parallel edges; this ensures that a minimum edge-cut in H does not contain any edges connecting the lower layer to the upper layer.) The same simple argumentation as above yields that EDGECUT has approximation ratio at most 2 for the min valid s-t-edge-cut problem. The proof of the following theorem, however, shows that this algorithm in fact computes an optimal solution.

Theorem 3. *There is a polynomial-time optimal algorithm for the min valid s-t-edge-cut problem in ToR graphs.*

Proof. We begin by proving a lemma that states a crucial property of (optimal) valid s-t-edge-cuts in ToR graphs.

Lemma 1. *Let $G = (V_G, E_G)$ be a ToR graph, $s, t \in V_G$ and C_G any valid s-t-edge-cut in G. From C_G an s-t-edge-cut C_H in the corresponding two-layer model $H = (V_H, E_H)$ can be derived with $|C_H| = |C_G|$.*

Proof. We start by adding for each edge $e \in C_G$ the two corresponding edges from the lower and upper layer to C_H. This yields an s-t-edge-cut C_H in H with $|C_H| = 2 \cdot |C_G|$, as already described in Section 3.1. Then, iteratively for each edge pair $e, e' \in C_H$ either e or e' is removed from C_H, where $e = (v, w) \in E_H$ is in the lower layer and $e' = (w', v') \in E_H$ is its counterpart in the upper layer. Below we show that assuming C_H is a cut before the removal it will still be a cut afterwards, if the edge is properly chosen. Thus, in the end, after considering all edge pairs in the original cut, C_H is still a cut and $|C_H| = |C_G|$ holds.

Now we consider a single step of the iteration where the pair $e = (v, w), e' = (w', v') \in C_H$ is treated, assuming that the (perhaps already modified) set C_H is still an s-t-edge-cut in H. Assume an s-t-path p exists that traverses only e and no other edge of the cut, i.e. $e \in p$ and $e_c \notin p$, for all $e_c \in C_H \setminus \{e\}$. We claim that in this case no s-t-path p' exists that traverses only e' and no other edge of the cut. It is then safe to remove e' from C_H. Symmetrically, if such a path p' exists, there cannot be a path p and thus e can be removed safely. (If neither p nor p' exists, remove e and continue the iteration.)

Aiming for a contradiction, we assume that both such paths p and p' exist. The edge $e \in p$ is directed from v to w, thus p has the form $s \cdots v \, w \cdots t$. Let $p_1 = s \cdots v$ denote the first part of p. Analogously the edge $e' \in p'$ is directed from w' to v' and thus p' has the form $s \cdots w' v' \cdots t$. Let $p_2 = v' \cdots t$ denote the last part of p'. Neither p_1 nor p_2 contain an edge from C_H. Therefore, p_1 and p_2 can be recombined via the edge (v, v') to form an s-t-path that does not contain any cut edge. This is a contradiction to the assumption that C_H is an s-t-edge-cut. □

Lemma 1 implies that the optimal s-t-edge-cut in H has at most as many edges as the optimal valid s-t-edge-cut in G. Conversely, every cut in H gives a valid cut in G of at most the same cardinality: consider the sets C_G and C_H in the Algorithm EDGECUT, clearly $|C_G| \leq |C_H|$ holds. Thus, an optimal s-t-edge-cut in the two-layer model H yields an optimal valid s-t-edge-cut in the ToR graph G. The former can be found in polynomial time by network flow techniques [1]. This concludes the proof of Theorem 3.   □

3.4   Max Vertex-/Edge-Disjoint Valid s-t-Paths

Theorem 4. *For a given ToR graph $G = (V, E)$ and $s, t \in V$, finding the maximum number of vertex- respectively edge-disjoint valid s-t paths is NP-hard. Moreover, the number of paths is even inapproximable within a factor $2 - \varepsilon$ for any $\varepsilon > 0$, unless P equals NP.*

Proof. We will reduce the problem of finding two disjoint paths between two pairs of terminals in a directed graph to this problem. Let G be a directed graph and s_1, t_1, s_2, t_2 four distinct vertices of G. Form a ToR graph G' from G by adding two vertices s and t, and edges from s to s_1, from t_1 to t, from t to s_2 and from t_2 to s. Note that no path $s, t_2, \cdots, t_1, t$ is valid. Thus, revealing the maximum number of vertex- respectively edge-disjoint valid s-t paths in G' would give a solution to the problem of finding two vertex- respectively edge-disjoint directed paths between s_1, t_1 and s_2, t_2 in G. The latter two problems are known to be NP-complete in general directed graphs [9].

   This also directly gives the inapproximability gap. For an arbitrary $k \in \mathbb{N}$, we simply make k copies of the graph G'. Next, we identify all copies of s to one node and all copies of t to one node. Intuitively, we now have k "parallel" copies of G. Depending on G there are either k or $2k$ vertex- respectively edge-disjoint valid paths between s and t. Let $\varepsilon > 0$ be some constant, independent of k. Clearly, if a $(2 - \varepsilon)$-approximation existed for max vertex- respectively edge-disjoint valid s-t paths, we could again solve the problem of finding two vertex- respectively edge-disjoint paths between s_1, t_1 and s_2, t_2 in G in polynomial time.   □

A Tight Approximation Algorithm. For ease of presentation we focus on the max vertex-disjoint valid s-t paths problem and comment at the end of the section how the result can be transfered to the max edge-disjoint case. In order to state the approximation algorithm, we need some definitions. If a forward part of a valid s-t path p_1 intersects the backward part of a path p_2 at a node v, we call this a *crossing* at v. The two paths can be *recombined* at the crossing to form a new path, consisting of the first part of p_1: $s, \cdots, v$ and the last part of p_2: $v, \cdots, t$. If p_1 and p_2 are recombined at v, the potential crossings on p_1 after node v and on p_2 before node v can be discarded. In the algorithm a recombination of such paths p_1 and p_2 may be revoked later on and thus not all these potential crossings can be discarded. However, it will be possible to discard the crossings of one of the paths. Note that p_1 and p_2 can be the same; in particular, if a path p contains a forward and a backward part, which meet at node u, we also say that the two parts cross at u.

ALGORITHM VERTEXDISJOINTPATHS

---

1. From G construct the two-layer model H, compute max vertex-disjoint s-t paths $\mathcal{P}_H$ in H.
2. Interpret $\mathcal{P}_H$ as set $\mathcal{P}_G$ of valid s-t paths in G. Note: $\mathcal{P}_G$ is not necessarily vertex-disjoint! Let $\mathcal{F}$ denote the forward parts of paths in $\mathcal{P}_G$ and $\mathcal{B}$ the backward parts. Recombine the parts as follows:
   (a) Select any not yet recombined forward part p_f in $\mathcal{F}$ that has at least one *remaining* (i.e. not discarded, see below) crossing.
   (b) Choose the first *remaining* crossing on p_f, let p_b in $\mathcal{B}$ be the corresponding backward part.
   (c) Recombine p_f and p_b, discard all previous crossings on p_b. In particular, if p_b was already recombined with p'_f, mark p'_f as not yet combined.
   (d) Repeat until each forward part is either recombined or has no remaining crossings.

Theorem 5. *The algorithm* VERTEXDISJOINTPATHS *is a 2-approximation algorithm for the max vertex-disjoint valid s-t paths problem.*

Proof. We first prove that the algorithm actually outputs a set of vertex-disjoint paths, then mention why the running time is polynomial and finally show that the approximation ratio of 2 is achieved.

Note that since the paths $\mathcal{P}_G$ are derived from vertex-disjoint s-t-paths in the two-layer model H, all forward parts $\mathcal{F}$ are disjoint and also all backward parts $\mathcal{B}$. Let $\mathcal{R}_i$ denote the set of recombined paths after the ith recombination. $\mathcal{R}_0$ is the empty set and thus vertex-disjoint. We now argue that if $\mathcal{R}_i$ is vertex-disjoint, then also $\mathcal{R}_{i+1}$ will be. In the $i+1$st recombination the selected forward part p_f does not intersect any backward part in $\mathcal{R}_i$ up to the chosen crossing, say at node v. Since step 2.(b) chooses the first remaining crossing on p_f, all potential crossings before v on p_f were discarded previously. We argue that also the rest of the backward part p_b: $v, \cdots, t$ does not intersect any other path q in $\mathcal{R}_i$. Assume the contrary, then there is a path q in $\mathcal{R}_i$ whose forward part q_f intersects p_b, say at node u. Let q_b be the backward part of q. Since the paths were derived from the two-layer model, at most two paths cross in each node. Thus q_f and q_b were recombined at a node $w \neq u$ and clearly w is after u on the forward part q_f (otherwise p_b would not intersect q_f). This gives the contradiction, since the algorithm would then have recombined q_f and p_b at node u instead of q_f and q_b at node w.

We have shown that the first part of p_f from s to v and the last part of p_b from v to t do not intersect any other path in $\mathcal{R}_i$. In case p_b was already recombined in $\mathcal{R}_i$ with some other forward part, this previous recombination is removed from $\mathcal{R}_i$, see step 2.(c). Thus, $\mathcal{R}_{i+1}$ is vertex-disjoint. As each crossing is considered at most once for recombination, the number of recombinations is $O(|V|)$ and the running time is polynomial.

It remains to prove the approximation ratio. Assume that k is the optimal number of paths for a given instance. Clearly $|\mathcal{P}_G|$ is at least k. For each path p in $\mathcal{P}_G$ either its forward part p_f, its backward part p_b, or both are recombined by the algorithm. If neither p_f nor p_b are recombined, p_f has at least one remaining crossing, namely the crossing with p_b. This crossing could not have been discarded, since p_b was never recombined with any forward part. Hence, either forward or backward part of each path are recombined, and thus at least $|\mathcal{P}_G|/2 \geq k/2$ disjoint valid s-t paths are found. $\square$

The algorithm can be easily adapted to the edge-disjoint paths setting. Here the crossings are at edges instead of nodes. The recombination of two paths that cross at an edge $e = (u, v)$ is done at node u, where e is directed from u to v. As in the computation of valid s-t-edge-cuts, $|V|$ parallel edges must be used to connect each vertex in the lower layer to its copy in the upper layer of the two-layer graph H; this ensures that any number of paths can switch from the lower layer to the upper layer at the same node. Analogously to the proof of Theorem 5 we obtain:

Theorem 6. *There is a 2-approximation algorithm for the max edge-disjoint valid s-t paths problem.*

3.5   On the Gap Between Disjoint Paths and Minimum Cuts

In the standard model of paths in directed or undirected graphs, Menger's theorem states that the maximum number of edge-disjoint s-t-paths is equal to the size of a minimum s-t-edge-cut, and the analogous result holds for vertex-disjoint paths and vertex-cuts (provided that there is no direct edge from s to t). As Menger's theorem applies to standard directed paths in the two-layer graph H, the proofs of Theorems 2 and 5 imply that for the valley-free path model there is always a valid s-t-vertex cut that is at most twice as large as the maximum number of vertex-disjoint valid s-t-paths. The same bound can be derived for the edge versions of the problems. Examples showing that the bound of 2 is tight are given in Fig. 3 both for the vertex version (left) and the edge version (right). In these examples, the size of a minimum valid s-t-cut is 2, while the maximum number of disjoint valid s-t-paths is 1.

Fig. 3. ToR graphs demonstrating a gap of 2 between disjoint paths and cuts

4   Max Vertex-/Edge-Disjoint Valid s-t-Paths in DAGs

We consider the problem of computing vertex- or edge-disjoint paths in directed acyclic graphs. This is motivated by the consideration that in a strictly hierarchical network, one would obtain ToR graphs that are acyclic. Due to space limitations, we refer the reader to [7] for proofs of the following results. First, we obtain that the problems remain *NP*-hard even for acyclic graphs.

Theorem 7. *For a given acyclic ToR graph $G = (V, E)$ and $s, t \in V$, finding the maximum number of vertex- respectively edge-disjoint valid s-t-paths is NP-hard.*

In general ToR graphs, it is *NP*-hard to decide whether there are two edge- or vertex-disjoint valid paths from s to t (Theorem 4). For acyclic graphs, we are able to show that this decision problem can be solved in polynomial time for any constant number of paths. Our proof is based on an extension of a pebbling game introduced by Fortune et al. [9].

Theorem 8. *For a given acyclic ToR graph $G = (V, E)$, $s, t \in V$, and any constant k, one can decide in polynomial time if there exist k vertex-disjoint (edge-disjoint) valid paths between s and t in G (and compute such paths if the answer is yes).*

5   Conclusions

We have initiated the study of disjoint valid s-t-paths and valid s-t-cuts in the valley-free path model. These problems arise in the analysis of the AS topology of the Internet if commonly used routing policies are taken into account. The size of a minimum valid s-t-vertex-cut can be viewed as a reasonable measure of the robustness of the Internet connection between ASs s and t. If the minimum cut has size k, this means that k ASs must fail in order to completely disconnect s and t. Therefore, our algorithms could be useful for network administrators who want to assess the quality of their network's connection to the Internet. Note that our approximation algorithm for the min valid s-t-vertex-cut problem can be easily adapted to the weighted version of the problem (where an AS that is unlikely to fail can be given a large weight).

The problems we have studied may be seen as instances of a more general family of problems whose common theme is that the allowed paths in the graph must obey certain restrictions. One example of such a restriction are oriented paths (paths containing at least one directed edge) in mixed graphs (graphs with undirected and directed edges), as considered by Wanke [18] in the context of the analysis of different parcellation schemes of the macaque brain. Another example are paths in graphs with labeled edges where a path is allowed only if the sequence of its edge labels forms a word from a given formal language; shortest-path problems for this type of restriction are studied by Barrett et al. [4] in the context of transportation problems. It would be interesting to study the max disjoint s-t-paths problem and min s-t-cut problem in such a setting.

There are also several open problems for the valley-free path model. It would be useful to study whether the maximum edge-disjoint or vertex-disjoint valid s-t-paths problem can be approximated better for acyclic graphs, and it would be interesting to determine the complexity and approximability of the min valid s-t-vertex-cut problem for acyclic graphs.

References

1. A. Ahuja, T. Magnanti, and J. Orlin. *Network Flows: Theory, Algorithms, and Applications.* Prentice-Hall, Englewood Cliffs, N.J., 1993.
2. G. Ausiello, P. Crescenzi, G. Gambosi, V. Kann, A. Marchetti-Spaccamela, and M. Protasi. *Complexity and Approximation. Combinatorial Optimization Problems and their Approximability Properties.* Springer, Berlin, 1999.
3. P. Baake and T. Wichmann. On the economics of Internet peering. *Netnomics*, 1(1), 1999.
4. C. L. Barrett, R. Jacob, and M. Marathe. Formal language constrained path problems. *SIAM J. Comput.*, 30(3):809–837, 2000.
5. E. Dahlhaus, D. Johnson, C. Papadimitriou, P. Seymour, and M. Yannakakis. The complexity of multiway cuts. *SIAM J. Comput.*, 4(23):864–894, 1994.
6. G. Di Battista, M. Patrignani, and M. Pizzonia. Computing the types of the relationships between autonomous systems. In *Proceedings of INFOCOM'03*, 2003.
7. T. Erlebach, A. Hall, A. Panconesi, and D. Vukadinović. Cuts and disjoint paths in the valley-free path model. TIK-Report 180, Computer Engineering and Networks Laboratory (TIK), ETH Zürich, 2003. Available electronically at ftp://ftp.tik.ee.ethz.ch/pub/publications/TIK-Report180.pdf.
8. T. Erlebach, A. Hall, and T. Schank. Classifying customer-provider relationships in the Internet. In *Proceedings of the IASTED International Conference on Communications and Computer Networks*, pages 538–545, 2002.
9. S. Fortune, J. Hopcroft, and J. Willie. The directed subgraph homeomorphism problem. *Theoretical Computer Science*, 10(2):111–121, 1980.
10. L. Gao. On inferring Autonomous System relationships in the Internet. *IEEE/ACM Transactions on Networking*, 9(6):733–745, 2001.
11. N. Garg, V. Vazirani, and M. Yannakakis. Multiway cuts in directed and node weighted graphs. In *Proceedings of ICALP'94*, LNCS 820, pages 487–498, 1994.
12. G. Huston. Interconnection, peering and settlements—Part I. *Internet Protocol Journal*, March 1999.
13. G. Huston. Interconnection, peering and settlements—Part II. *Internet Protocol Journal*, June 1999.
14. C. Labovitz, A. Ahuja, R. Wattenhofer, and S. Venkatachary. The impact of Internet policy and topology on delayed routing convergence. In *Proceedings of INFOCOM'01*, 2001.
15. L. Subramanian, S. Agarwal, J. Rexford, and R. Katz. Characterizing the Interenet hierarchy from multiple vantage points. In *Proceedings of INFOCOM'02*, 2002.
16. H. Tangmunarunkit, R. Govindan, and S. Shenker. Internet path inflation due to policy routing. In *Proceedings of SPIE ITCom'01*, 2001.
17. H. Tangmunarunkit, R. Govindan, S. Shenker, and D. Estrin. The impact of routing policy on Internet paths. In *Proceedings of INFOCOM'01*, 2001.
18. E. Wanke. The complexity of finding oriented paths in mixed graphs. Technical report, Institut für Informatik, Heinrich-Heine-Universität, Düsseldorf, 2003.

A Distributed Algorithm to Find Hamiltonian Cycles in $\mathcal{G}(n,p)$ Random Graphs*

Eythan Levy[1], Guy Louchard[1], and Jordi Petit[2]

[1] Département d'Informatique, Université Libre de Bruxelles,
Bld du Triomphe — CP 212, B-1050 Bruxelles, Belgium
{elevy, louchard}@ulb.ac.be
[2] Departament de Llenguatges i Sistemes Informàtics, Universitat Politècnica de Catalunya, Campus Nord Ω-227. 08034 Barcelona, Catalonia
jpetit@lsi.upc.edu

Abstract. In this paper, we present a distributed algorithm to find Hamiltonian cycles in $\mathcal{G}(n,p)$ graphs. The algorithm works in a synchronous distributed setting. It finds a Hamiltonian cycle in $\mathcal{G}(n,p)$ with high probability when $p = \omega(\sqrt{\log n}/n^{1/4})$, and terminates in linear worst-case number of pulses, and in expected $O(n^{\frac{3}{4}+\epsilon})$ pulses. The algorithm requires, in each node of the network, only $O(n)$ space and $O(n)$ internal instructions.

1   Introduction

Many random models of graphs have been recently proposed and studied as models of computer networks (see e.g. [14, 13] for random Web graph models and [6] for a random graph model designed for optical networks of sensors). In such distributed networks, resolving graph-theoretic problems whose instance is given by the topology of the network can have interesting applications. A classical and well-studied problem of this type is the distributed minimum spanning tree [2, 7]. In this paper, we study another well known graph theoretic problem in such a distributed context : the Hamiltonian cycle problem. We propose a distributed heuristic algorithm that searches for a Hamiltonian cycle in the graph induced by the topology of the network, and analyze its efficiency in terms of probability of success and time complexity, when the topology is given by the classical $\mathcal{G}(n,p)$ model of random graphs.

The Hamiltonian Cycle Problem. A Hamiltonian cycle is a cycle that visits each vertex of a graph exactly once. If a graph has a Hamiltonian cycle, it is said to be Hamiltonian. It is well known that deciding whether a graph is Hamiltonian is an **NP**-complete problem. One classical way to deal with such hard problems

---

* This research was partially supported by the EU within the 6th Framework Programme under contract 001907 (DELIS) and by the Spanish CICYT project TIC2002-04498-C05-03 (TRACER).

A. López-Ortiz and A. Hamel (Eds.): CAAN 2004, LNCS 3405, pp. 63–74, 2005.

is to devise polynomial-time heuristic algorithms and show that their probability of failure is low, or even asymptotically null, under some probability distribution of the inputs.

$\mathcal{G}(n, p)$ *Random Graphs.* In this paper we use the well-known $\mathcal{G}(n, p)$ random graph model of Erdös and Renyi. In this model, graphs contain n vertices labeled $\{1, \ldots, n\}$ and each of the $\binom{n}{2}$ possible edges are independently included with probability p. The probability p can be taken as a constant, in which case the average number of edges is $p\binom{n}{2} = O(n^2)$ and a dense graph results or it can be defined as a decreasing function p_n of n, which produces sparser graphs on average (fixing $p_n = \log n/n$ for example yields graphs with an average number of edges proportional to $n \log n$). A closely related model is the $\mathcal{G}(n, m)$ random graphs model, which consists of graphs of exactly n nodes and m vertices, each one of these graphs having equal probability. Most of the time, the two models are practically interchangeable, provided m is close to p/n (see [3] for a classical reference).

A large collection of results on $\mathcal{G}(n, p)$ random graphs are available, among which many *threshold* results, that express the minimum density required to have a certain property with high probability (whp). In particular, for any divergent function $t(n)$, a graph in $\mathcal{G}(n, p_n)$ is Hamiltonian whp for $p_n = (\log n + \log \log n + t(n))/n$; see [3].

Hamiltonian Cycle Heuristics. Several heuristic algorithms have been proposed to deliver whp Hamiltonian cycles in $\mathcal{G}(n, p_n)$ graphs provided that p_n is sufficiently large. These algorithms return a Hamiltonian cycle if they succeed in finding one, they otherwise return that no Hamiltonian cycle has been found. They are heuristic in the sense that they might not find any Hamiltonian cycle though one may exist. [1] devised an $O\left(n \log^2 n\right)$ algorithm to find Hamiltonian cycles w.h.p in $\mathcal{G}(n, m)$ random graphs when $m > cn \log n$ for some constant c. [10] presented a linear time algorithm for finding a Hamiltonian path in $\mathcal{G}(n, p)$ graphs with constant p. The HAM algorithm in [4] finds Hamiltonian cycles w.h.p in $\mathcal{G}(n, m)$ random graphs when $m = n \log n/2 + n \log \log n/2 + t(n)n$, with $t(n) \rightarrow \infty$ and runs in $O\left(n^{4+\epsilon}\right)$ time. It is essentially best possible with regards to the density of the graph. Finally, [20] gives a $O(n/p_n)$ algorithm to find Hamiltonian paths in $\mathcal{G}(n, p_n)$ random graphs when $p_n \geq 12n^{-\frac{1}{3}}$.

On the other hand, it has also been shown that there exist exact algorithms that produce a Hamiltonian cycle in a graph or establish the nonexistence of such a cycle, and run in polynomial expected time over $\mathcal{G}(n, p)$. These algorithms proceed by combining a heuristic algorithm with an exponential exact one, applying the latter only if the former fails to find a Hamiltonian cycle in the graph, and showing that the probability of success of the heuristic algorithm is high enough as to render the contribution of the exponential algorithm to the average complexity negligible. [4] show that an heuristic algorithm, combined with an exact $O\left(n^2 2^n\right)$ dynamic programming algorithm works in polynomial expected time when $p \geq \frac{1}{2}$. [10] also combine two heuristic algorithms and an exact one to solve the Hamiltonian Path problem in expected linear time when p is a constant. Finally, [20] obtains a more general result by combining two heuristic algorithms and an exact one to solve the Hamiltonian Path problem in expected $O(n/p_n)$ time when $p_n \geq 12n^{-1/3}$.

All the above cited algorithms were designed for classical sequential computers. Some exact algorithms for finding Hamiltonian cycles in $\mathcal{G}(n, p)$ on parallel computers have been proposed: Frieze [9] proposed two algorithms for EREW-PRAM machines: the first uses $O(n \log n)$ processors and $O(\log^2 n)$ time, while the second one uses $O(n \log^2 n)$ processors and $O((\log \log n)^2)$ time. MacKenzie and Stout [17] proposed an algorithm for Arbitrary CRCW-PRAM machines that operates in $O(\log^* n)$ average time and requires $n/\log^* n$ processors. All these parallel algorithms assume p is a constant.

All the algorithms we have cited above are designed either for RAM or PRAMs machines. However, to the best of our knowledge, a fully distributed algorithm for this problem has not been yet proposed. By "fully distributed algorithm", we mean an algorithm designed for a network of computers connected according to a certain topology, and in which the computers have disjoint memories and communicate only through message-passing between *direct* neighbors.

More precisely, we desire a fully distributed algorithm that, when executed on the nodes of a network of computers whose topology is given by a certain graph, ends in a state where: either a Hamiltonian cycle has been found and all the nodes of the network end up knowing their two neighbors in this cycle, either the algorithm fails to find such a cycle, and all nodes end up knowing of this failure.

Distributed Computation Model. Our algorithm works in the classical model of synchronous networks (see e.g. Chapter 12 in [19]), where the algorithm takes place in a sequence of discrete steps, called *pulses*, in which every process first sends (zero or more) messages, then receives all the messages addressed to it during that same pulse, and finally performs local computations. Communication is limited to the direct neighbors. The *time complexity* of an algorithm under this model is defined as the number of pulses needed for the algorithm to terminate, while the *message complexity* is defined as the total number of messages exchanged during the execution of the algorithm. We further suppose that the nodes know the identities of their direct neighbors, that a fixed initiator node, say v_0, knows the size (number of nodes) of the network initially, and finally that the nodes have an inner memory whose size is $O(n)$. If we allowed the presence of a memory of $O(n^2)$ in the nodes, then our Hamiltonian cycle problem could be trivially solved in time $O(diameter)$ by collecting the entire graph topology in one node, and having that node compute a Hamiltonian cycle locally and broadcast the result to the other nodes.

Notice that our model of distributed computation is less powerful than a PRAM machine in the sense that in a parallel machine, all the processors have access to a centralized shared memory, while in our case the memories are disjoint. This has the important implication that none of the nodes in our network has access to the total adjacency matrix of the graph underlying the network: the distributed processors must collaborate in order to find a Hamiltonian cycle in a graph with an adjacency matrix of which every processor knows only one line.

Applications. A fully distributed Hamiltonian cycle algorithm has interesting applications in the fields of distributed computation. For instance, with a Hamiltonian cycle it is possible to build a path to perform distributed computations based on end-to-end communication protocols, which allow distributed algorithms to treat an unreliable network as a reliable channel [18]. Also, a Hamiltonian cycle is useful for the purpose of forming token rings in the network, establishing a sense of direction, and as part of distributed algorithms for election or mutual exclusion [19]. Our algorithm can also find useful applications in emerging systems, as we will discuss in the conclusions.

Summary of Results. In this paper, we present and analyze a fully distributed randomized heuristic Hamiltonian cycle algorithm for $\mathcal{G}(n,p)$ random graphs. Our focus in this paper is on the time complexity and the probabilistic properties of our algorithm, we do not attempt to optimize message complexity (recall that the optimization of time and message complexities are often conflicting goals in distributed computation). The algorithm is designed for $\mathcal{G}(n,p)$ graphs in the sense that the graph underlying the topology of the network is a $\mathcal{G}(n,p)$ graph, and the nodes are labeled in a way consistent with this graph.

Within this setting, the paper is organized as follows: First, we present a high level description of the distributed algorithm. Then, we analyze its probability of success over the probability space of $\mathcal{G}(n,p_n)$ graphs for a suitable probability function p_n. Our results show that w.h.p. our algorithm finds Hamiltonian cycles when $p_n = \omega\left(\sqrt{\log n}/n^{1/4}\right)$. Finally, we prove that the worst-case and average time complexities of our algorithm are respectively $O(n)$ and $O\left(n^{3/4+\epsilon}\right)$, and that the algorithm requires only linear space and a linear total number of internal computation steps in each node of the network. We close the paper with some concluding remarks.

Due to lack of space, we omit a formal exposition of the algorithm and simply sketch most of the proofs; see [16] for details.

2   High Level Description of the Algorithm

Our algorithm follows the same general working scheme as MacKenzie and Stout's algorithm [17] and works in three main sequential phases:

1. *Initial cycle phase:* In this phase an initial small cycle with $\Theta\left(\sqrt{n}\right)$ vertices is found in the graph.
2. *Path covering phase:* In this phase, almost all the vertices out of the initial cycle are covered by $\sqrt{n}$ vertex-disjoint paths.
3. *Patching phase:* In this phase, each path and each non covered vertex is patched into the initial cycle.

Let us give some more details on each phase. Figure 1 depicts the phases of the algorithm.

Phase 1 (Initial Cycle). The goal of this phase is to build an initial cycle of length $\Theta\left(\sqrt{n}\right)$. In order to build such a cycle, the algorithm proceeds in two

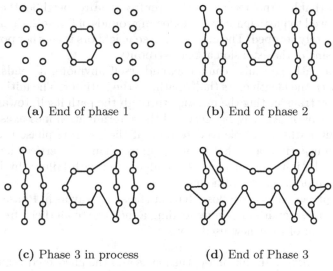

(a) End of phase 1   (b) End of phase 2

(c) Phase 3 in process   (d) End of Phase 3

Fig. 1. Phases of the algorithm

steps. First, it sequentially builds a path of length $\lambda_1 = 6\sqrt{n}$ beginning with the initiator vertex v_0; then, it tries to loop this path back to v_0 by extending the path as long as the path extremity is not adjacent to v_0. The algorithm stops its execution with a failure —and broadcasts the failure information to all nodes— if the length of the path becomes greater than $\lambda_2 = 7\sqrt{n}$ or if it does not succeed in extending the path. Broadcasting in a graph of arbitrary topology can be done in linear worst-case time with respect to the diameter using a wave algorithm [19].

Vertices in the cycle (and in the paths of Phase 2) are said to be "used" and vertices out of the cycle are said to be "free". At all times the nodes should know the subset of their neighbors that are still free. To achieve this, the new endpoint of the current path always sends a message to all its (free) neighbors, notifying them of its transition to the "used" state.

Phase 2 (Path Covering). The goal of this phase is to cover almost all the vertices not included in the cycle by a set of $\sqrt{n}$ vertex-disjoint paths leaving, at most, $\sqrt{n}$ vertices uncovered.

In order to cover the vertices, the $\sqrt{n}$ paths will grow in parallel from a set of $\sqrt{n}$ initial vertices chosen by v_0 among its free neighbors, after the completion of the initial cycle. A path extends its-self by its two extremities in the following way: an extremity chooses one of its free neighbors uniformly at random, and sends an extension message to it, waiting for its answer. These choices are synchronized between all the participating extremities.

Free vertices wait for extension messages emanating from the extremities and pick one of these messages uniformly at random, to which they answer. Then, each one of these free neighbors thus becomes the new extremity of one of the paths, and will execute this same extension mechanism at the next round of Phase 2. All path extremities that have not received an answer from the free neighbor they had chosen are not allowed to participate to the following rounds of Phase 2; their extension is then completely stopped. The absence of free neighbors is another possible reason for the ceasing of the extension at one extremity.

When a path extremity cannot extend itself anymore, it sends its identity, together with the length of its (half-)path, to the initiator. The initiator is always reachable by transmitting the message through the path itself, toward the initial node of that path. After v_0 has received these termination messages from all the $\sqrt{n}$ covering paths, it is able to determine if the covering phase is successful or not. In the negative case, that is, when more than $\sqrt{n}$ vertices have remained uncovered, the initiator terminates the algorithm with failure, by broadcasting the failure information to all nodes.

In this phase, we also suppose that, in the same way as in Phase 1, the newly selected path extremities begin by sending a message to all their (free) neighbors, notifying them of their new used state.

Phase 3 (Patching). In this phase, the paths and the uncovered vertices are tried to be patched to the initial cycle. In the case that all of them can be patched to the cycle, a Hamiltonian cycle will be returned; otherwise the algorithm will report failure. In the following, uncovered vertices will be treated as paths of length zero.

Phase 3 starts by gathering in v_0 the identities of the uncovered vertices. This can be done in time $O(diameter)$ using a wave algorithm as shown in [19].

Patching an individual path to the cycle is done according to the simple idea depicted in Fig. 2: if u and v are two consecutive vertices in the cycle and s and t are the two endpoints of the path, the path can be patched to the cycle if edges us and vt or ut and vs exist in the graph. Patching a path with length zero to a cycle is done in the same way, just taking $s = t$.

In practice, the patching trials, for a fixed path, are done using a patching message that circulates round the cycle, and contains the identities of s and t, as well as two boolean variables denoting whether the sender of the message is adjacent to s and t. Let $C_1, \ldots, C_k$, with $C_1 = v_0$, be the nodes of the cycle. The message is initially launched by v_0, and upon arrival at a node C_k, checks whether the path is patchable to nodes C_{k-1} and C_k. If it is the case, then nodes s, t, and C_{k-1} are notified of the cycle update and the patching of the path terminates. If not, the patching message is updated, and sent by C_k toward C_{k+1}. If the patching message loops back to the initiator with no success, then the patching for that path has failed.

The overall patching of the paths is done in parallel, by pipelining several patching messages —one per path— on the cycle. These messages are initially launched by v_0, separated by a delay of three time pulses, in order to avoid possible inconsistencies in the patchings performed by two adjacent messages. This delay between the messages explains the constants 6 and 7 used in Phase 1: they guarantee that the cycle is long enough to contain all the patching messages. When a patching trial succeeds, a notifying message is sent backward toward the initiator, along the cycle. Thus, at the end of Phase 3, the initiator knows exactly how many patching trials have succeeded, and performs one last broadcast in order to notify all the nodes of the final result of the algorithm, being success or failure. In the successful case, all nodes have the knowledge of their two neighbors in the Hamiltonian cycle.

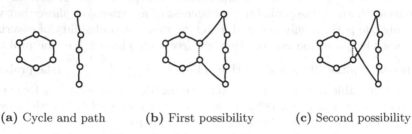

(a) Cycle and path   (b) First possibility   (c) Second possibility

Fig. 2. Patching a path into a cycle

3   Analysis of the Probability of Success

In this section, we show that the algorithm succeeds w.h.p. in finding a Hamiltonian cycle in a $\mathcal{G}(n,p)$ random graph when $p_n = \omega\left(n^{-1/4}\sqrt{\log n}\right)$. Let I_n^i be the indicator random variables denoting the success of Phase i conditioned to the success of Phase j, with $j < i$. Then, denote by $I_n = I_n^1 \cdot I_n^2 \cdot I_n^3$ the indicator random variables denoting that the algorithm finds a Hamiltonian cycle.

Lemma 1. *For each $\epsilon > 0$, we have:* $\boldsymbol{Pr}[I_n^1 = 1] \to 1$ *when* $p_n \geq n^{-\frac{1}{2}+\epsilon}$.
Sketch of the proof. A simple way to prove the lemma is to show that the following variant algorithm (which has trivially a lower probability of success than ours) succeeds w.h.p. for the demanded p_n: first compute a path of length $\lambda = 7\sqrt{n}$ (phase 1a), then then try to close it by successively trying the nodes (backward), doing a maximum number of $\sqrt{n}$ trials (phase 1b).

The probability of success of phase 1a can easily be shown to be $\prod_{i=1}^{\lambda}(1 - (1-p_n)^{n-i})$, which is greater than $\left(1 - (1 - p_n)^{n-7\sqrt{n}}\right)^{7\sqrt{n}}$. This last expression can now be shown to tend to 1 using the classical asymptotic approximation techniques of the exp-log transformation and the asymptotic equivalence: $(1 - \epsilon_n)^{f_n} \sim e^{-\epsilon_n f_n}$ when $\epsilon_n \to 0$ and $f_n \to \infty$.

Concerning phase 1b, since $\sqrt{n}$ nodes are tried for closing the cycle, we obtain $1-(1-p_n)^{\Theta(\sqrt{n})}$ for its probability of success. This probability can also be shown to tend to 1.

The probability of success of phase 1 is now the product of the probabilities of success of phases 1a and 1b, and hence tends to 1.   □

Lemma 2. *For each $\epsilon > 0$, we have:* $\boldsymbol{Pr}[I_n^2 = 1] \to 1$ *when* $p_n \geq n^{-\frac{1}{2}+\epsilon}$.
Sketch of the proof. In order for phase 2 to succeed, the initiator must first find $\sqrt{n}$ free neighbors in order to start the extension (phase 2a), and then launch the extension itself (phase 2b).

Let l be the actual length of the initial cycle. The number of free neighbors of the initiator is given by a $B(n - l, p_n)$ binomial random variable (r.v.). Using $l \geq 6\sqrt{n}$ and Chebychev's inequality, it is easy to show that w.h.p., this r.v. is greater than $\sqrt{n}$ when $p_n \geq n^{-\frac{1}{2}+\epsilon}$.

Concerning phase 2b, we bound (below) the probability of success of our extension phase by the probability of success of an extension phase that would use only one path. If only one path extends, then the probability of covering all but possibly $\sqrt{n}$ of the vertices that are free after phase 1 (i.e. the probability of success of phase 2b) is greater than $\left(1 - (1 - p_n)^{\sqrt{n}}\right)^{n-l-\sqrt{n}}$ (the probability of not being able to extend the temporary path when at least $\sqrt{n}$ free vertices remain is lower than $(1-p_n)^{\sqrt{n}}$ and we must do at least $n-l-\sqrt{n}$ such extensions in order to succeed). The above expression tends to 1.

The probability of success of phase 2 is now the product of the probabilities of success of phases 2a and 2b, and hence tends to 1.   □

Lemma 3. *For all $p_n = \omega\left(n^{-1/4}\sqrt{\log n}\right)$, $\boldsymbol{Pr}[I_n^3 = 1] \to 1$.*

Sketch of the proof. For the sake of simplifying the proof, we consider a simpler variant patching algorithm whose probability of success is trivially lower than ours, and show that this probability tends to 1. This variant tries, alternatively, each other edge for patching, which yields independent patching trials.

The probability of being able to patch a path at a *fixed* position on the cycle is $\geq p_n^2$ (it is p_n^2 if the path has length 0 and $2p_n^2 - p_n^4$ otherwise). The probability of not being able to patch a path *anywhere* on the cycle is $\leq \left(1 - p_n^2\right)^{\sqrt{n}}$ (the number of patching trials —half the length of the initial cycle— is always higher than $\sqrt{n}$).

The probability of being able to patch all $\sqrt{n}$ paths is therefore greater than $\left(1 - \left(1 - p_n^2\right)^{\sqrt{n}}\right)^{\sqrt{n}}$, which tends to 1.   □

From the preceding results, we obtain:

Theorem 1. *For all $p_n = \omega\left(n^{-1/4}\sqrt{\log n}\right)$, we have $\boldsymbol{Pr}[I_n = 1] \to 1$.*

4   Complexity Analysis

In this section, we perform a complexity analysis of our algorithm and show that its worst-case time and space complexities, as well as the total number of instructions executed in each node are linear in n. We also study the best and average-case time complexities of our algorithm and show the former to be $\Theta(\sqrt{n})$ and the later to be sub-linear, namely $O(n^{\frac{3}{4}+\epsilon})$ for each $\epsilon > 0$.

Our first theorem concerns the worst-case complexities:

Theorem 2. *The worst-case time complexity of the algorithm is $O(n)$, as are the worst-case space needed in each node, and the worst-case total number of instructions executed in each node.*

Proof. The time complexity of phase 1 is trivially bounded by $7\sqrt{n}$. . The time complexity of phase 2 is bounded by that of the worst possible scenario, in which none but one of the paths succeeds in growing at each pulse. In that case, only one free node is covered at each pulse, yielding a worst-case time bounded by $O(n)$. Finally, concerning phase 3, the time complexity of the patching of each path is linear at worse, since the patching message circulates at worst once completely on the cycle. Since the patchings of the paths are done in parallel and the time interval between the starts of the patching trials of the first and last paths is $O(\sqrt{n})$, we obtain a bound of $O(n)$ for the total time complexity of phase 3. We still have not counted the time complexity of the final broadcasting wave. A wave algorithm can be executed in time $O(diameter)$ (see chapter 6 of [19]), which is $O(n)$.

Concerning the space complexity, the only data structure our nodes need to use is a list of free neighbors, whose size is trivially linear.

Finally, the total number of local instructions performed by the nodes is $O(n)$. This number is bounded by the worst-case number of deletions from its free neighbors list that a node must make, which is $O(n)$.   □

Our second theorem concerns the best-case time complexity:

Theorem 3. *The best-case time complexity of the algorithm is $\Theta(\sqrt{n})$.*

Proof. Phase 1 always terminates in time $\Theta(\sqrt{n})$, phase 2 covers all nodes in time $\Theta(\sqrt{n})$ when all the paths succeed in growing by one unit until all free nodes are covered, and finally phase 3 terminates in the same time bound when the patching of each path succeeds at the first try. Concerning the final broadcasting wave, its time complexity id bounded by $O(\sqrt{n})$ whenever the diameter of the graph is bounded by that quantity.   □

We now come to the average-case time complexity analysis. Let T_n be the random variable denoting the time complexity of our algorithm on a $\mathcal{G}(n, p_n)$ random graph, with $p_n = \omega\left(n^{-1/4}\sqrt{\log n}\right)$. Also, let T_n^1, T_n^2, T_n^3 and T_n^4 be the random variables denoting respectively, the costs of Phases 1, 2, 3 and of the termination wave.

The average-case time complexity of phase 1 is trivially $\Theta(\sqrt{n})$:

Lemma 4. $\mathbf{E}[T_n^1] = O(\sqrt{n})$.

Concerning phase 2, we are able to obtain a bound of $O\left(n^{3/4+\epsilon}\right)$:

Lemma 5. *For all $\epsilon > 0$, $\mathbf{E}[T_n^2] = O\left(n^{3/4+\epsilon}\right)$.*

Sketch of the proof. The proof is based on a random urns model for the extension phase, in which we redefine the free nodes as urns, and the path endpoints as balls being thrown at random in the urns; see Fig. 3. We then analyze this urns model using advanced probabilistic tools such as limiting theorems for urns occupations and Brownian motion.   □

Let b_i and u_i denote, respectively, the number of balls and urns at round i

$i := 0$

repeat until $b_i = 0$ or $u_i = 0$:
    for each ball:
        With probability $(1 - p)^{u_i}$, discard the ball
        if ball not discarded:
            Throw the ball in a uniform random urn
    for each non-empty urn:
        Discard the urn
        Keep one of the balls of the urn for next round
        Discard all other balls of the urn
    $i := i + 1$

Fig. 3. The extension phase seen as an urns model

Lemma 6. *Let S_m be the supreme of m independent and identically distributed geometric variables having parameter $p = 2m^{-1} \ln m$. We have:*

$$\mathbf{E}[S_m] \sim \frac{m}{2} \left(1 + \frac{\gamma}{\ln m}\right), \quad \text{where } \gamma \text{ is Euler's constant.}$$

Sketch of the proof. In short, proceeding as in [11], we prove the convergence of S_m to a Gumbel distribution (a distribution having distribution function $F(x) = e^{-e^{-x}}$), we compute the rate of convergence, and we analyze the convergence of moments. □

We now characterize the average time complexity of Phase 3:

Lemma 7. $\mathbf{E}[T_n^3] = O(\sqrt{n})$.
Sketch of the proof. In short, we bound the complexity of our patching procedure of a path to the cycle by the complexity of an alternative patching procedure that only tries, alternatively, each other edge for patching. This yields independent patching trials, and permits us to bound the time needed to patch a path to the cycle by two times a geometric random variable with parameter p_n^2. Since the patchings of the different paths are done in parallel, we are looking for the mean of the supreme of $\sqrt{n}$ of these r.v.'s and use lemma 6 to this end. □

Concerning the wave executed at the beginning of phase 3 and the final broadcast wave, their average time cost is bounded by $O(\mathbf{E}[diameter])$. The diameter of a $\mathcal{G}(n, p_n)$ random graph is known to be constant w.h.p. for graphs as dense as ours (see [3]), but we are rather searching for the expected diameter. We obtain the following lemma:

Lemma 8. $\mathbf{E}[T_n^4] = O(\log n/p_n) = o\left(n^{1/4}\sqrt{\log n}\right)$.
Sketch of the proof. We bound the diameter of the graph by the height of a random tree inspired by the Galton–Watson process outlined in [12]. and then show the average height of that tree to be $O(\log n/p_n)$ using the saddle point method (see [8]). □

Our main theorem follows now from Lemmata 4, 5, 7 and 8:

Theorem 4. *Let T_n be the random variable denoting the execution time of the distributed algorithm when $p_n = \omega\left(n^{-1/4}\sqrt{\log n}\right)$. Then, for all $\epsilon > 0$, $\mathbf{E}[T_n] = O\left(n^{3/4+\epsilon}\right)$.*

5   Conclusion

In this paper we have presented a randomized distributed algorithm to find Hamiltonian cycles of graphs. Its analysis on the standard model of $\mathcal{G}(n, p_n)$ random graphs with $p_n = \omega(\sqrt{\log n}/n^{1/4})$ shows that the algorithm delivers Hamiltonian cycles with high probability and that its expected running time is sub-linear. Also, in each node both the total number of computation steps performed and the total space required are not unreasonable: linear at most, which could be useful for implementations on networks of low-cost devices.

Many distributed algorithms have been proposed to cope with graph theoretic problems. However, only a few studies have concentrated on a probabilistic analysis of these algorithms, namely in the aspects of average-case complexity, the probability of success of heuristic algorithms, and randomized network topologies. This new kind of results may be of use in the design and analysis of the emerging global systems resulting from the integration of autonomous interacting entities, faulty or dynamic links and ad-hoc mobile networks where wireless and mobile networks have a dominating role. For instance, the algorithm we have proposed can be used in order to get a distributed solution to find Hamiltonian cycles in random geometric networks with edge faults, which can model sensor networks, as $\mathcal{G}(n, p_n)$ random graphs arise naturally as subgraphs of such networks (see [5] and [15]).

The hypothesis of synchrony in our network model can be seen as a strong constraint. It should be noted however (see [15]) that the algorithm, under some minor modifications, can be run in an asynchronous network. In such a setting, our results on high probability of success and worst-case complexities of the algorithm remain valid, though our average-case complexity results no longer hold.

Acknowledgments. The authors would like to thank S. Langerman, J. Cardinal, C. Lavault and J. Díaz for their precious comments.

References

1. D. Angluin and L. G. Valiant. Fast probabilistic algorithms for hamiltonian circuits and matchings. *Journal of Computer and System Sciences*, 18:155–193, 1979.
2. B. Awerbuch. Optimal distributed algorithms for minimum-weight spanning tree, counting, leader election and related problems. In *Proc. 19th Symp. on Theory of Computing*, pages 230–240, 1987.
3. B. Bollobás. *Random graphs*. Academic Press, London, second edition, 2001.

4. B. Bollobás, T. I. Fenner, and A. M. Frieze. An algorithm for finding Hamilton paths and cycles in random graphs. *Combinatorica*, 7(4):327–341, 1987.
5. J. Díaz, J. Petit, and M. Serna. Faulty random geometric networks. *Parallel Processing Letters*, 10(4):343–357, 2001.
6. J. Díaz, J. Petit, and M. Serna. A random graph model for optical smart dust networks. *IEEE Transactions on Mobile Computing*, 2(3):186–196,, 2003.
7. M. Faloutsos and M. Molle. Optimal distributed algorithm for minimum spanning trees revisited. In *Symposium on Principles of Distributed Computing*, pages 231–237, 1995.
8. P. Flajolet and R. Sedgewick. The average case analysis of algorithms: Saddle point asymptotics. Technical Report RR-2376, INRIA, 1994.
9. A. Frieze. Parallel algorithms for finding hamilton cycles in random graphs. *Information Processing Letters*, 25:111–117, 1987.
10. Y. Gurevich and S. Shelah. Expected computation time for hamiltonian path problem. *SIAM Journal on Computing*, 16(3):486–502, 1987.
11. P. Hitczenko and G. Louchard. Distinctness of compositions of an integer: A probabilistic analysis. *Random Structures & Algorithms*, 19(3-4):407–437, 2001.
12. S. Janson, T. Łuczak, and A. Rucinski. *Random graphs*. Wiley, New York, 2000.
13. J. M. Kleinberg, R. Kumar, P. Raghavan, S. Rajagopalan, and A. S. Tomkins. The Web as a graph: Measurements, models and methods. *Lecture Notes in Computer Science*, 1627, 1999.
14. S. R. Kumar, P. Raghavan, S. Rajagopalan, and A. Tomkins. Extracting large-scale knowledge bases from the web. In *VLDB Journal*, pages 639–650, 1999.
15. E. Levy. Distributed algorithms for finding hamilton cycles in faulty random geometric graphs. Mémoire de licence (master's thesis), Université Libre de Bruxelles, http://www.ulb.ac.be/di/scsi/elevy/, 2002.
16. E. Levy. Analyse et conception d'un algorithme de cycle hamiltonien pour graphes aléatoires du type $g(n,p)$. Mémoire de DEA, Ecole Polytechnique, Paris, http://www.ulb.ac.be/di/scsi/elevy/, 2003.
17. P. D. MacKenzie and Q. F. Stout. Optimal parallel construction of hamiltonian cycles and spanning trees in random graphs. In *ACM Symposium on Parallel Algorithms and Architectures*, pages 224–229, 1993.
18. S. Nikoletseas and P. Spirakis. Efficient communication establishment in adverse communication environments. In J. Rolim, editor, *ICALP Workshops*, volume 8 of *Proceedings in Informatics*, pages 215–226. Carleton Scientific, 2000.
19. G. Tel. *Introduction to Distributed Algorithms*. Cambridge University Press, second edition, 2000.
20. A. G. Thomason. A simple linear expected time algorithm for finding a hamilton path. *Discrete Mathematics*, 75:373–379, 1989.

String Matching on the Internet*

Hervé Brönnimann, Nasir Memon,
and Kulesh Shanmugasundaram

Polytechnic University,
Six Metrotech, Brooklyn, NY 11201, USA
{hbr, memon}@poly.edu, kulesh@cis.poly.edu

1 Introduction

We consider a variant of the "string searching in database" problem where the string database comes on a data stream, and processing the data is at a premium but querying is not a runtime bottleneck. Specifically, the strings to be searched into (let's call them the documents) have to be processed online very efficiently, meaning the documents have to be added to some string searching data structure one by one in time proportional to their length. Of course, we desire this data structure to be small, i.e. at most linear space, and hopefully exhibit a tradeoff between storage/processing cost and accuracy. Upon some query string, the data structure must return whether that string is contained in a document (the *presence query*), and must also be able to return a list of the documents which contain the query (the *attribution query*). We may require that the query be large enough and that only portions of it may match (pattern matching). In practice, it is acceptable that the data structure return a superset of the answer, as long as no document from the answer is missing and there are only few false positives; either the false positives can be filtered (by actual verification if the document texts are available in a repository), or a small number of false positives are acceptable for the application (e.g. network forensics, see below).

The prototypical framework which motivates this research is where documents are network packets which have to be processed at line speed in a network node. The data structure must be stored in the node and thus be as compact as possible. The application we have in mind is *network forensics*: we would like to be able to trace back criminal activity based on content information (worm or virus signature, spam excerpt, etc.). By placing a few such nodes in our networks, we record enough information to ask which packets (including source and destination, time frame, etc.) have carried certain portions of payload. While there are some traceback techniques which work with packet header information, we believe we are the first to propose content-based attribution, as a first step to traceback. The principal challenge is to process the packets at line speed.

---

* This research is supported by NSF CyberTrust Grant 0430444, "Fornet: :Design and Implementation of a Network Forensics System".

A. López-Ortiz and A. Hamel (Eds.): CAAN 2004, LNCS 3405, pp. 75–89, 2005.

In principle, the query can take all the time it wants, as long as it has a good accuracy (is able to perform attribution with a small number of false positives).

This framework may also have applications in string matching on data streams, and for computational biology of sequences, where approximate string matching and online constructions play an important role. Note that the meaning of approximate here is different from those linked to edit-distance, and have to do with the low number of false positives and the fact that only portions of the query string may be matched.

Our contribution. We propose a hash-based method for solving the problem above. Namely, we decompose the input documents into blocks (using fixed-size blocks or value-based block), and use Bloom filters to hash these blocks into a bit vector which can later be queried. However, as such, this method (which we call *Block Bloom filters*, or *BBF*) requires a rather large space, does not provide a very high accuracy, and only answers presence queries. Moreover, it is prone to a number of intrisic false positives when the documents can be controlled by an adversary (for example, blocks identical to the query blocks but not in consecutive positions, or even query blocks in consecutive positions but in different documents).

The novel ideas which we introduce to bolster the accuracy are to construct a hierarchy of blocks of different lengths (*Hierarchical Bloom Filter*, or *HBF* for short), and engineer it for the attribution problem. In order to make sure the blocks occur in consecutive position, we also hash the block offset in the document with the block and later query the Bloom filter with all possible offsets (in our application, the documents are short and the largest offset is not very large). In order to perform attribution queries, we also hash the document ID with the blocks and later query the Bloom filter with every possible document ID if the presence query is positive. The accuracy of this method degrades with the number of documents (which is very high in our application), and yet we show it to remain very good in practice. Intuitively, the hierarchy takes care of blocks which do not occur consecutively, or occur consecutively but in different documents, and hence drastically lowers the false positive rates. We provide an analysis that the HBF outperforms the BBF, and some experimental evidence that the HBF is very efficient at attributing query strings to documents, even when the number of documents is very large.

Note: This workshop paper is part of a whole system we are building for network forensics. We discuss the overall architecture of the system in [14] and discuss the payload attribution problem and the engineering of HBFs in [13], along with sharper experimental results, as well as real use cases. Portions of sections 2 and 3 are present in that paper as well. This paper differs from [13] in that it presents the analysis of HBF vs. BBF and discusses Rabin fingerprinting for value-based blocking (not in [13]), but avoids the discussion of payload attribution and related issues (such as resistance against various attacks, privacy, etc.; the interested reader should refer to [13]). In short, this paper focuses on the algorithmic content, and provides all the analysis we could not provide in the conference submission for reasons of space. Some of the experimental results also overlap but are provided here in abstracted form for completeness.

Related work. The string searching in database problem has a host of applications, including web caching and proxying, searching an email/text database, matching mitochondrial DNA for identification, detecting plagiarism in a collection of documents, and in an on-line scenario similar to the one we consider here, eliminating network redundancy and deep packet inspection. In an off-line context, the problem is usually solved by building an inverted index (when the queries are words whose order don't matter, and may not even appear consecutively). This is the standard situation in web searching, and there results are usually ordered (using various ranking heuristics, most famously *pagerank*, and various algorithms like Fagin's algorithm to merge the ranked lists). When we would like to match substrings instead of words, a popular technique is to build a search structure on the text, such as a suffix tree or suffix arrays. This is the standard situation in sequence processing (e.g. computational biology). Unfortunately, those structures are memory- and time-consuming, and somewhat involved especially to construct in linear time online. Furthermore, it appears that memory paging is a problem and this cannot be afforded in our application where disk I/O is much slower than packet processing speeds. Reducing storage can be achieved by grouping the input string into words, and making the words the basic unit in the matching. Even the best implementations we are aware of take at least 9~12 bytes of whatever we choose to make the alphabet, and don't allow pattern matching. For these reasons, we decided not to explore suffix trees and arrays further, although we do not know whether the best online constructions of suffix arrays would permit the processing rates and storage we achieve here.

2   Hierarchical Bloom Filters

2.1   Basic Bloom Filters

A Bloom filter [1] is a simple, space-efficient, randomized data structure for representing a set in order to support membership queries. It uses a set of k hash functions of range m and a bit vector of length m. Initially, the bit vector is set to 0. An element in the set is inserted into the Bloom filter by hashing the element using the k hash functions and setting the corresponding bits in the bit vector to 1. The space efficiency of a Bloom filter is achieved at the cost of a small probability of false positives as defined by Equation 1, where n is the number of elements in the set [3].

$$FP = \left(1 - (1 - \frac{1}{m})^{kn}\right)^k \approx (1 - e^{-kn/m})^k. \tag{1}$$

To optimize the FP rate above, the following relationships between the parameters should be respected: $k = ln2 \cdot (m/n)$, for which $FP \approx 0.6185^{m/n}$.

Many variants of Bloom filters have been proposed. Spectral Bloom filters [5] extend the data structure to support estimates of frequencies, which is also the goal of Space Code Bloom filters [9]. Counting Bloom filters [8] enable deletions as well as insertions. When the filter is intended to be passed as a message, compressed Bloom filters [11] may be used with worse raw storage but better compressed size. Finally, the most recent variant is Bloomier filters [4] which allow to associate an r-bit value per element of the set instead of just storing a set: the value stored is always correct if the element is in the set, and a small number of false positives will have a bogus associated value.

Bloom filters have received a lot of attention for networking applications (see the recent survey [2]). In recent work Bloom filters have been used in the Source Path Isolation Engine (SPIE) to trace IP packets over networks [15]. Finally, very recently Dharmapurikar et al. [7] propose to use Bloom filters for intrusion detection by matching known signatures in the payload. Their system uses very similar principles (blocks of different length, all possible alignments) but proceeds on the other end (the Bloom filter contains the signatures) and for a different purpose (intrusion detection).

2.2   Block-Based Bloom Filter (BBF)

A shortcoming of the suffix tree-based methods is that they allow to match a query string anywhere in the documents, exactly. This necessitates at least linear storage [6]. A well-used technique for trading off storage with granularity is to block the input. The most immediate way to do this is by splitting the documents into blocks of fixed size s bytes. The larger s is, the smaller the storage. But queries of length less than s bytes cannot be answered, and even for queries longer than s bytes, up to $2(s-1)$ at each end may have to be dropped. Moreover, since we do not know at what offset the query will be in the document, when querying all decompositions of the query with blocks of size s must be tried (the first block having any number of bytes between 0 and $s-1$).

The basic BBF technique is to insert every of the blocks into a single Bloom filter. Upon query, the Bloom filter replies if the block is present in a document or not. In order to reply positively for a query, all the query blocks have be found in the filter. Note that this does not allow to find which document that is. (Bloomier filters could be used for that but with much higher storage since they would have to associate a document ID to each block, instead of a single bit.) It also suffers from a few drawbacks we discuss in the next section. Nevertheless, it will be the basis for our construction. We call this a *Block-Based Bloom Filter (BBF)*.

Another well-known technique decomposes the documents into blocks of variable size whose boundaries are dictated by the content. This technique, which uses Rabin fingerprints, was pioneered by Manber [10], and applied in networking [16] and web caching [12]. The idea is to compute a running hash function on p consecutive bytes of the input, and declare a block boundary when the last $r = \lfloor \log_2 s \rfloor$ bits of the hash are 0 (which happens with probability $1/2^r \approx 1/s$ if the hash is random enough and leads to blocks of expected size $\leq s$). The hash function of choice is a Rabin fingerprint which can be computed with only few operations per additional byte in the input. The same hash can be applied to the query, resulting in the same block decomposition and therefore save us from trying every possible alignment of the query inside the document.

Unfortunately, there is no guarantee for the longest nor the shortest block. If one so desires, one may enforce a minimum and maximum block size, at the cost of the block boundaries not depending on the input value only (we call those boundaries *artificial* to emphasize that they don't depend on the input). In fact, enforcing size requirements may cascade and force articial boundaries all throughout (when maximum limit is reached every single time, if value-based boundaries occur before minimum is reached after the artificial boundaries). This is highly unlikely and the value-based boundaries are expected to reappear shortly after artificial boundaries, but the possibility cannot be discounted and thus may force us to have to try every possible alignment of the query again (unless we are willing to tolerate a few very unlikely *false negative* answers to our query). There is one way to get rid of this phenomenon: to enforce only a maximum limit but no minimum (at the potential cost of high storage if boundaries happen too often, which would be a deviation of the random behavior of the hash function).

In conclusion, value-based blocking will help if we do not care about small or large blocks, or if we can tolerate a few very unlikely *false negative* answers to our query. If we do care only about large blocks (and can live with very small blocks), we may fix only a maximum size limit for blocks and still reap the benefit of value-based caching. Fixed-size blocks offer somewhat more predictable storage and computation costs, but at the cost of more expensive (and potentially less accurate) queries. In our forensics scenario, this is not a big problem, and we cannot however accomodate large blocks nor false negatives (that would hamper the usefulness of a forensic query), so we chose fixed-size blocks. But in other settings, value-based blocking would yield superior results which is why we mention it here.

2.3   Hierarchical Bloom Filter (HBF)

Applying BBF for string queries suffers from the following drawbacks which we call collisions. Namely, if we query for a string whose blocks are present in several documents (some blocks in one, others in another, etc.) then we are led to believe that the query is present in some document whereas it is not. This is not due to the randomization in Bloom filters, and will occur no matter what the hash functions are. It can be somewhat avoided by using random-hash value-based blocking, but nevertheless, a long query whose first half is in a document and second half (overlapping substantially with the first) in another will exhibit this problem, no matter how random the Rabin fingerprint is chosen. One possible answer is to hash the offset along with the block (both in the insertion and in the query, where all block numbers have to be tried in sequence). In order to reply positive, all query blocks have to test positive at consecutive offsets. But even that could be fooled if the excerpts happen at indeed consecutive offsets in different documents.

We circumvent both these possibilities by hashing roughly twice as many blocks, but introducing some redundancy in the blocks by coallescing two consecutive blocks of the BBF calling this a level-1 block. In general, blocks at level i are paired two by two and coallesced into a level-$(i + 1)$ block. Every block of the hierarchy is inserted into the Bloom filter. We call this a *Hierarchical Bloom Filter (HBF)*. Since there are twice as many blocks to insert into the Bloom filter, to retain the same storage as BBF, we must either use twice a bigger block size, or with the same block size as the BBF the basic false positive rate must be $FP_{HBF} = \sqrt{FP_{BBF}}$. This will be our assumption for the upcoming analytical comparison.

With the preceding scheme, the boundaries of blocks at any level are a subset of the boundaries of the underlying BBF (level 0). An intriguing possibility arising with value-based blocking is to base the level boundaries to achieve a geometric evolution of the block lengths (level i with expected length $2^i s$) but to use different block criteria per levels (e.g. the last $i + \log_2 s$ bits of the Rabin fingerprint form a predetermined bit pattern b_i such that b_i is not a prefix nor a suffix of b_{i+1}). We have not experimented with this approach nor can we offer an analysis.

2.4   Adapting HBF for Answering Attribution Queries

In this section we describe in detail how to answer attribution queries using a HBF. Note that the construction of BBF and HBF described above can only verify whether a string queried was seen by the HBF or not. However, if we would like to attribute a string to a document then content must be tied to a particular document. This can easily be accomplished by inserting an additional substring of the form (content‖offset‖docID) for each block inserted into HBF, where *docID* is assigned online and/or stored in a different table as documents keep coming. In our application, it could be a string that identifies the host that originated or received the payload, for instance, SourceIP, DestinationIP, and/or port information. If the documents are stored in a database, the primary key of the document or simply a sequence number can be used. During attribution the attribution system need a list of candidate *docID*s from which it can choose a set of possible attributions. This may either be supplied by the query, or stored as part of the HBF itself during the construction.

Overall, the system we propose is organized in several tiers, as depicted in Figure 1. For each incoming document, the system produces the blocks in the hierarchical block decomposition of the document. For every such block, the information $(content, offset, docID)$ is available, and the system maintains:

1. a **block digest** (optional): a HBF storing the hashes of (content).
2. an **offset digest**: a HBF storing the hashes of (content‖offset).
3. a **document digest**: a HBF storing, for every block (content‖offset) in the offset digest, the corresponding (content‖offset‖docID).

One may use a BBF instead of an HBF but HBF leads to fewer false positives. The main advantage of using a block digest is to have better accuracy

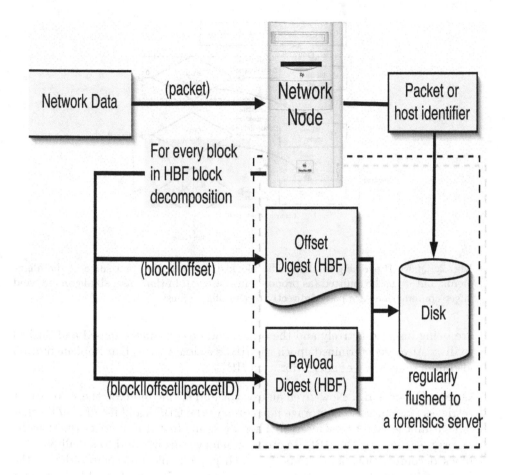

Fig. 1. A high level view of the system, with emphasis on packet processing and HBF. The optional block digest is not represented

at answering whether a block has been seen at all (without knowing the offset). Without it, one must query the offset digest with all possible offsets: although the extra space afforded by not having a block digest increases the accuracy of the offset digest, the testing of every offset gives both designs roughly equivalent accuracy. So, we can omit the block digest and save storage to increase the accuracy of the offset digest. Nevertheless, if there are lots of queries for small excerpts, it may be beneficial to keep a block digest.

Input Processing. Based on the rate of arrival of documents and the required accuracy of attribution, appropriate HBF parameters for accuracy and space are determined a priori. The system maintains two HBFs, one into which documents

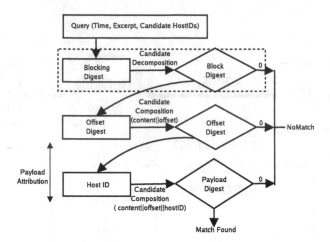

Fig. 2. Query Processing in HBF. The block digest will filter out some of the alignments, but it can be omitted (as proposed in the text). In that case, all alignments and offset combinations are passed directly to the offset digest

are being inserted actively and the other ready to be time-stamped and flushed to disk. At a predetermined epoch the HBFs switch places. Our implementation also maintains a list of $docID$ for each HBF.

Query Processing. Now given an excerpt and a time interval, the system first retrieves (from persistent storage if necessary) the HBF's and list of $docIDs$ that fall within the target time interval. Then we would first like to verify whether the excerpt was seen by the HBF. In order to achieve this we need to try all possible block decompositions and offsets. For each possible block decomposition of the input string, verify the existence of every content. If any of the blocks cannot be found, then the query string has not been seen by the HBF. If every single block is verified by the HBF, then we need to make sure they appear in the same order as in the query string. To verify the order, we append all possible offsets to the strings (content‖offset) and verify their positions. Based on their offset we may be able to go to a higher level in the HBF hierarchy and increase the confidence as described earlier. Now, in order to attribute the query string we simply append the $docIDs$ from the list being maintained by our PAS for the particular HBF being queried and verify the (content‖offset‖docID). The combinations of docID+alignment Figure 2 depicts how a query is processed in such a setup.

2.5　Extensions

One may extend the above system to process queries that span multiple documents. For instance, in our network setting, a content may be split into several packets. One may process those queries by testing, if a block is not present, for all possible *early terminations* and resume with the remainder of the query blocked from that point on, with an offset 0 and a different *docID*.

Another extension concerns a limited form of pattern matching. It is possible to ask queries of the form xy (i.e. "retrieve all documents which contain a copy of x, followed later by a copy of y"), and in fact any combination. The query can easily be answered by taking the intersection of all the documents that contain x and those that contain y, but the order relationship can also be determined by comparing the offsets that match x and y within those documents. One must keep in mind that small string at both ends of both x and y may not be matched due to the granularity of the HBF.

3   Analysis of BBF and HBF

Given a string x of length p, whose blocks are $x_1 \ldots, x_{p/s}$ at offsets $i_1, \ldots, i_{p/s}$, if the string belongs to a document, then it will be definitely be found by a query since Bloom filters have no false negatives. Hence the only parameter to evaluate is the false positive rate of the overall query (we call this the *effective FP rate*). In the analysis below, we will compare the results when using a standard BBF or a HBF in the digests.

In the analysis, we will use the worst case of fixed-size blocking, for two reasons: 1. it is the one we have implemented for the reasons described in section 2.2, and 2. the analysis is valid for any block decomposition scheme, with the provision that it can knock off a factor s from the FP_e in case of value-based blocking.

Formulation of the problem, and parameters. In the following, we try and obtain estimates for the effective FP rate for a given query. Remember that we may have up to three Bloom filters for storing (content), (content||offset) or (content||offset||docID). We use the following notation:

FP_e = effective FP rate for payload query (after all the filtering)
FP_o = individual FP rate of (content) and (content||offset) BF for BBF or HBF,
FP_p = individual FP rate of (content||offset||docID) BF for BBF or HBF
 n_b = number of blocks inserted in the BBF,
 n_h = number of blocks inserted in the HBF ($n_b \leq n_h \leq 2n_b$),
 N = the number of *docID*s to check for in the document digest.

Note that $FP_p = FP_o$ if all standard BFs are merged together. Other parameters, such as the size of the Bloom filters, can be determined from these parameters. For instance the size of the BBF, m_b, satisfies $FP_o = 0.6185^{m_b/n_b}$, and the number of hash functions used in the BBF is $\ln 2 \cdot (m_b/n_b)$.

3.1   Analysis of BBF False Positives

In order for the document digest to give a positive answer, all the digest (block, offset and document digests) must give a positive answer. A false positive for a query with $\lfloor p/s \rfloor$ blocks can be obtained for several reasons (and combinations thereof), outlined in Section 2.

For the analysis, let us use u to represent the number of blocks of the query which do not belong to the input or which do appear but at the wrong offset (over all possible offsets of the query string), which we call *false blocks*, and v the minimum number of blocks which do appear at the right offset but with the wrong packet ID (over all the possible packet IDs). Note that these parameters depend on the query and on the set of documents for which the query is a positive. For a false positive, in the worst case, $u = 0$ and $v = 1$ (all blocks are present and all but one are present in the same document), but by expressing FP_e as a function of (u, v) we gain more insight in the performance of our data structure.

The number of possible block decompositions is s for each possible block alignment, but there are only $\lfloor L/s \rfloor$ possible offsets, where L is the maximum length of a document. This induces $s\lfloor L/s \rfloor \approx L$ possible candidates. The u false blocks must fool the BBF. Hence we expect to see false positives pass after the (content||offset) BF with probability at most $L \cdot FP_o^u$.

Note: The above assumes that we did not use a block digest as suggested in Section 2.4. If we did, the above probability would be at most $L \cdot FP_b'^u FP_o'^u$. Nevertheless, since the space savings of not having a block digest enables FP_o to be about $FP_b' FP_o'$, it turns out both solutions have about the same accuracy.

To prevent offset collisions, since we must test N packet IDs for each passing combination of blocks with offsets, we expect to see a false positive after both (content||offset) and (content||offset||docID) BFs with probability at most

$$FP_e^{(BBF)} = L \cdot N \cdot FP_b^u \cdot FP_p^{u+v}.$$

Typically N is very large, and it may thus be necessary to lower the rate FP_p of the standard BF to keep FP_e within a reasonable range. This also shows that the effective rate FP_e crucially depends on the relation between the query and the input documents (the parameters u, v are unknown to the analyst, barring some kind of probabilistic model on the input and query.)

3.2   Extension to HBF

When using HBF instead of BBF, with an equal memory footprint m, the basic FP rate is higher, i.e., $FP_o' \approx \sqrt{FP_o}$, and $FP_p' =\approx \sqrt{FP_p}$ since the HBF stores approximately $n_h \approx 2n_b$ blocks (perhaps fewer due to number of blocks per packet not being always powers of two). All the u false blocks and false (block,offset) pairs are contained in another u' blocks and pairs in the hierarchy, and v false (content||offset||docID) triplets are contained in another v' triplets in the hierarchy.

The basic observation is this: *block aggregates in the hierarchy that contain a false block must be false as well.* In the worst-case, the false blocks are consecutive and their ancestor form a subtree in the hierarchy, with $u' = u-1$ internal nodes. But this could be much more advantageous: If there are l levels to the hierarchy, we could have up to $u' \approx lu$. Moreover, two blocks which happen consecutively but not in a single document will not be inserted as (content||offset) in the HBF at a higher level. Therefore the hierarchy will also catch a subset of the offset collisions (those that occur together in the hierarchy). Let u'' be the number of aggregates in the HBFs which correspond to such collisions. (We expect that $u'' \approx v$, although due to rounding powers of two, this may not be exact.) Thus we expect a negative query to pass the (content||offset) HBF with probability

$$FP_e^{(HBF)} = L \cdot (FP_b')^{u+u'+u''}.$$

Since there is no need to involve docIDs to resolve offset collisions, thanks to the hierarchy, we obtain a much better false positive rate, that does not involve N at all.

When we do involve docIDs for packet attribution, the false positive rate after querying both (content||offset) and (content||offset||docID) with all possible offsets and docIDs is still better. Indeed, we expect to see a false positive with probability at most

$$L \cdot N \cdot (FP_o')^{u+u'+u''} (FP_p')^{u+u'+u''+v+v'}.$$

Note that u'' and v bear similar relationship and hence, with equal memory requirements, since $u + u' \geq 2u - 1$ and $FP_o' \approx \sqrt{FP_o}$, clearly $FP_e^{(HBF)} \leq FP_e^{(BBF)}$, with potential for $\ll$ instead of $\leq$.

As a conclusion, with equal memory requirements, HBF is never worse than BBF, and sometimes much better.

4   Experimental Results

In this section, we focus on the specific network forensic setting and analyze how the effective FP rate for payload query depends on the parameters considered so far (block size s, basic FP rate for the various Bloom filters). In particular we must describe how to choose the basic payload digest parameters in order to achieve a good FP rate for the overall payload query, both for BBF and HBF.

For this purpose we used a packet trace that comprises all email-related traffic (IMAP, SMTP, and POP3) between $1,500$ hosts for a 24-hour period. The trace is about 1.5GB and contains approximately 3.3 million packets. Although this is not a very big volume of data, it is enough to estimate the quantities involved and the results would scale with larger data sets. Also, the results are independent from the nature of the traffic.

4.1   Collision Rate of Payloads

One the parameters of our document database which we first set to measure is the conditional probability P of offset collision. The rationale is that knowledge of the collision rate allows us to choose a good block size for the filters and might also clue us in on which block size s and memory requirements to choose

Table 1. Probabilities of offset collisions by protocol, measured on a network trace of 1M packets. Note: these are *not* the false positive rates of our filters, see Table 3 for those

Query Blocks	SMTP	HTTP	FTP
16	0.542259	0.562018	0.168082
32	0.762793	0.547919	0.050411
64	0.410678	0.362524	0.006253
128	0.313077	0.322957	0.003306
256	0.213254	0.282814	0.003717

to obtain acceptable FP_e. As it turns out, this probability yields much too pessimistic conclusions, and we will discuss why at the end of this subsection. The choice of parameters in the following subsections will not be based on this analysis, yet it is interesting to report our findings here.

Measuring the conditional probability P of offset collision can be achieved by sorting for every block position i, the documents according to the values X and Y of their blocks i and $i + 1$ (lexicographically). Let $N_{i,X}$ be the number of documents whose ith block is X, and $N_{i,X,Y}$ the number of those whose next block is Y. The number of pairs of packets sharing the same block at some position is $\sum_{i,X} N_{i,X}{}^2$, and among those pairs, those that share the same block at the next position is $\sum_{i,X,Y} N_{i,X,Y}{}^2$. Note that we take squares and not binomials $\binom{N}{2}$ since we want to count the pairs of two identical packets. Those can be computed directly by sorting the trace lexicographically and repeatedly for every value of i. Hence

$$P = \frac{\sum_{i,X,Y} N_{i,X,Y}{}^2}{\sum_{i,X} N_{i,X}{}^2},$$

which is measured experimentally for various block sizes and types of network traffic.

Table 1 lists the collision rates against the block sizes for three commonly used protocols. Collision rate is measured over one million packets from each trace.

Surprisingly, we find that this probability is quite high even for moderate block sizes. The reason is very likely that most of the traffic is highly structured. For example, notice that the probability is very high for SMTP, HTTP but relatively lower for FTP. This is because, SMTP commands and HTTP headers are highly structured and tend to repeat frequently (automatic mail checking and auto refresh), increasing the regularity of the trace. FTP, on the other hand, is least structured as it transfers data in bulk with very few commands and doesn't repeat very often reducing the regularity in the trace. For highly structured protocols, it would be interesting to segregate the payloads into two kinds (headers or application data) and measure P separately for the two parts. One could also arrive at a prediction for P following the entropy of the English language.

The numbers we obtain in this section suggest that there are many collisions, hence that our solution may not be feasible. In fact, we will prove otherwise

Table 2. Effects of block perturbation of query string on the false positive rate of a
HBF with base false positive rate $FB_o = 0.0109$

Perturbed Blocks	1	2	3	4
	0.111237	0.002011	0.002305	0.000032

in the subsequent section. The reason they don't apply to our system, is that
the actual FP rate of a payload query depends on the query string itself. If this
string does not include the SMTP commands that drive P high, the analysis
does not apply and a much lower FP_e can be expected. How low is the subject
of the next experiments.

Next, we measured the impact of the query string structure on the effective
false positive rates. For this experiment, we chose 50,000 query strings of size
5 blocks each, that are in fact in the HBF. Special care was taken in choosing
the strings such that they had the worst-case alignment in the HBF. Then we
perturbed the blocks in the query string (1 through 4) and measured the false
positive rate (See Table 2).

Clearly the strings with fewer blocks being perturbed have higher false positive rates than those with multiple blocks. The query strings being in worst-case
alignment also contribute to the increase in false positive as some of the blocks
can only be verified at a single level in the hierarchy. For the forensics analyst, this has the implication that the results is more accurate if the excerpt is
"unique" or at least has the most distinctive features. Portions which can be
repeated in many payloads do not bring a good improvement to the effective
rate FP_e.

4.2   Experimental Evaluation of the Effective FP Rate FP_e

The results in this section have been reported in our conference submission
and so we only report the table of results and refer the interested reader to our
other paper for the full discussion. Nevertheless, we present here one short table.
Table 3 presents the effective FP rate (as measured) of HBF, as a function of
FP_o and s. The results speak for themselves.

Table 3. Measured effective false positive rate (FP_e) of HBF as a function of both
the basic false positive rate (FP_o) and the length of the query (in blocks; 1block=32
bytes). Note that for $blocks > 4$, we encountered no false positives, hence the measured
FP_e is equal to 0 (indicated by –)

Blocks	Basic False Positive Rates (FP_o)							
	0.3930	0.2370	0.1550	0.1090	0.0804	0.0618	0.0489	0.0397
1	1.000000	0.999885	0.996099	0.976179	0.933179	0.870477	0.798657	0.728207
2	0.063758	0.064569	0.048981	0.036060	0.026212	0.021024	0.015881	0.012538
3	0.012081	0.002620	0.000744	0.000275	0.000172	0.000046	0.000023	–
4	0.000820	0.000230	0.000060	0.000020	–	–	–	–
> 4	–	–	–	–	–	–	–	–

5   Conclusion and Future Work

In this paper, we introduce the problem of string matching in a stream of network packets. Although we were motivated by IP networks, the data structure we propose can be adapted to work in several other situations where we search in a string database that is presented on a data stream, and where false positives can be dealt with. Biology-motivated sequence problems, for instance, may benefit from such speedups.

The system we have described is part of a larger system for facilitating network forensics across and within networks. The system we have implemented monitors network traffic, creates hash-based digests of payload, and archives them periodically. A user-friendly query mechanism provides the interface to answer post-mortem questions about the payload. Hardware implementation and feasibility of a payload attribution system on high-speed networks are part of our future work.

References

1. B. Bloom. Space/time tradeoffs in hash coding with allowable errors. In *Communnications of the ACM* 13(7):422-426, 1970.
2. A. Broder and M. Mitzenmatcher. Network applications of Bloom filters: A survey. In *Annual Allerton Conference on Communication, Control, and Computing*, pages 636-646, 2002.
3. P. Cao. Bloom filters - the math. http://www.cs.wisc.edu/ cao/papers/summary-cache/node8.html.
4. B. Chazelle, J. Kilian, R. Rubinfeld, and A. Tal. The Bloomier filter: An efficient data structure for static support lookup tables. In *Proc. ACM/SIAM Symposium on Discrete Algorithms*, pages 30–39, 2004.
5. S. Cohen and Y. Matias. Spectral Bloom filters. In *Proc. ACM SIGMOD International Conference on Management of Data*, pages 241–252, 2003.
6. E. D. Demaine and A. Lopez-Ortiz. A linear lower bound on index size for text retrieval. *Journal of Algorithms* 48(1):2–15, 2003. Special issue of selected papers from the 12th Annual ACM-SIAM Symposium on Discrete Algorithms (SODA 2001).
7. S. Dharmapurikar, M. Attig, and J. Lockwood. Design and implementation of a string matching system for network intrusion detection using fpga-based bloom filters. Technical Report, CSE Dept, Washington University, 2004. Saint Louis, MO.
8. L. Fan, P. Cao, J. Almeida, and A. Z. Broder. Summary cache: A scalable wide-area web cache sharing protocol. *IEEE /ACM Transactions on Networking*, 8(3):281-293, 2000.
9. Abhishek Kumar, Li Li, and Jia Wang. Space-code bloom filter for efficient traffic flow measurement. In *Proc. of the Conference on Internet Measurement*, pages 167–172, Miami Beach, FL, USA, 2003.
10. U. Manber. Finding similar files in a large file system. In *Proc. of the Winter 1994 USENIX Conference*, pages 1–10, San Francisco, CA, 1994.
11. M. Mitzenmacher. Compressed Bloom filters. *IEEE/ACM Transactions on Networking*, 10(5):613-620, 2002.
12. S. C. Rhea, K. Liang, and E. Brewer. Value-based web caching. In *Proc. 12th International Conference on World Wide Web*, pages 619–628. ACM Press, 2003.

13. K. Shanmugasundaram, H. Brönnimann, and N. Memon. Payload attribution via hierarchical bloom filters. In *Proc. of the ACM Conference on Computer Communications and Security*, pages 31–41, 2004.

14. K. Shanmugasundaram, N. Memon, A. Savant, and H. Brönnimann. Fornet: A distributed forensics network. In *Proc. of MMM-ACNS Workshop*, pages 1–16, 2003.

15. A. C. Snoeren, C. Partridge, L. A. Sanchez, C. E. Jones, F. Tchakountio, S. T. Kent, and W. T. Strayer. Single-packet IP traceback. *IEEE/ACM Transactions on Networking* 10(6):721–734, 2002.

16. N. T. Spring and D. Wetherall. A protocol-independent technique for eliminating redundant network traffic. In *Proc. of the Conference on Applications, Technologies, Architectures, and Protocols for Computer Communication*, pages 87–95. ACM Press, 2000.

k-Robust Single-Message Transmission

Extended Abstract

André Kündgen[1], Michael J. Pelsmajer[2], and Radhika Ramamurthi[3]

[1] Department of Mathematics,
California State University, San Marcos
akundgen@csusm.edu
[2] Department of Applied Mathematics,
Illinois Institute of Technology,
pelsmajer@iit.edu
[3] Department of Mathematics,
California State University, San Marcos
ramamurt@csusm.edu

Abstract. End-to-end communication considers the problem of sending messages between a sender s and a receiver r through an asynchronous, unreliable network, such as the Internet. We consider the problem of transmitting a single message from s to r through a network in which edges may fail and cannot recover. We assume that some s, r-path survives, but we do not know which path it is. A routing algorithm is k-robust if it ensures that a message sent by s will be received by r when at most k edges fail, and it will never generate an infinite number of messages. Graphs with a k-robust algorithm for all k were characterized in [5]. For any other graph, its *robustness* is the maximum k for which it has a k-robust algorithm.

We provide general lower bounds for robustness by improving a natural algorithm obtained from Menger's Theorem. We determine robustness for several examples, such as complete graphs, grids, and hypercubes.

1 Introduction

End-to-end communication considers the problem of sending messages between a sender s and a receiver r through an asynchronous, unreliable network, such as the Internet. We consider the simplified problem of transmitting a single message through a network in which edges may fail and cannot recover. Making no assumptions about the speed at which a message travels along an edge, an edge that has failed is indistinguishable from an edge along which packets are traveling very slowly. Therefore, a solution cannot use information about which edges have failed. In public networks, intermediate processors do not store information about the state of the communication between s and r, and our model conforms to this restriction. Thus, when a processor receives a packet, it must immediately decide to which of its neighbors to send packets, basing its decision only on the edge along which the packet arrived. We seek a protocol that ensures

A. López-Ortiz and A. Hamel (Eds.): CAAN 2004, LNCS 3405, pp. 90–101, 2005.

that a message sent by s will be received by r if not too many edges fail. We could guarantee this by simply having each processor forward a copy of each arriving packet along all incident edges; however, this would usually have the consequence of r being flooded with copies of the message. To avoid this problem we seek algorithms that will send only finitely many copies of a message to r, no matter which edges fail.

How do we measure the quality of such a routing algorithm? We say that such an algorithm is *k*-robust if a message sent from s is guaranteed to be received at r whenever (i) at most k edges fail, and (ii) for some s, r-path, none of its edges fail. Thus, if there is any hope of sending a message from s to r, then the algorithm will do so.

Networks that permit a routing algorithm which is *k*-robust for all k were characterized in [5] by a list of 10 forbidden minors. These infinitely robust graphs are structurally very similar to outerplanar graphs. Lovász [4] observed that there are relatively few such networks because there is no limit on the number of edges that can fail, and he suggested studying the case when "few" edges fail. This leads us to the following definition: The *robustness* of a network is the maximum k for which it has a *k*-robust algorithm. The purpose of this paper is to study this notion.

We begin in Section 2 by translating this discussion into graph theoretic terms and proving some basic equivalences. A generalization of infinite robustness which allows fixed-size packet headers that contain routing information is studied in [1, 2, 5], and descriptions of related models and results can be found in [1, 3]. In particular, [1] considers protocols for complete graphs and [2] considers protocols for $m \times n$ meshes (grids). We determine the robustness of these graphs in Section 4, as well as for hypercubes and the Möbius ladder of order 8. This last graph is particularly interesting for two reasons. One, it gives us an example for which an optimal algorithm must send messages "backward" in a sense which doesn't occur in optimal algorithms for meshes, hypercubes, and complete graphs. Two, the graph has a symmetry which fixes s and r, and yet no optimal routing algorithm may have that symmetry; for our other examples we give optimal routing algorithms that behave nicely which are intuitive with respect to the structure of the graphs. In Section 3 we obtain a sharp general lower bound in terms of the edge-connectivity between s and r. This result improves upon a natural routing algorithm obtained from Menger's Theorem.

2   Definitions and Notation

Our network model is similar to the one used in [5], and we describe it as follows.

In this paper, we always consider a connected graph with two distinct special vertices s (sender) and r (receiver). An *instruction* is a list evf such that v is a vertex that is incident to edges e and f. If $e = uv$ and $f = vw$ and there are no multiple edges, we can also denote this by uvw. A routing algorithm A is a set of instructions.

Given a message, A begins by sending it along each edge incident to s. Then a message arriving at v from edge e is sent along f whenever evf is an instruction.

A *routing* is a walk $v_0, e_1, v_1, \ldots, e_k, v_k$ in which all $e_i v_i e_{i+1}$ for $1 \leq i \leq k - 1$ are instructions. We also denote it by $v_0, v_1, \ldots, v_k$ if that will not cause confusion, and refer to it as a v_0, v_k-routing. Note that a routing algorithm will send a message from s to r if and only if there is an s, r-routing. To avoid having r flooded with copies of the message, we seek algorithms with only finitely many s, r-routings.

Next, we describe conditions equivalent to having only finitely many s, r-routings. We say that an instruction is *essential* for a routing algorithm if it is contained in an s, r-routing. For routing algorithms A and A', A' is a *strengthening* of A if every essential instruction of A is also an essential instruction of A'. A sequence of vertices $v_0, \ldots, v_k$ with $v_0 = v_k$ is an *essential cycle* if there are edges $e_0, \ldots, e_k$ with $e_0 = e_k$ such that $e_{i-1} v_i e_i$ is an essential instruction for $1 \leq i \leq k$.

A labeling l of a graph is an assignment of two numbers $l(u, v)$ and $l(v, u)$ for each edge $e = uv$. Note that this can be adapted to multigraphs with slightly different notation. The *routing algorithm derived from l* consists of all instructions of the form uvw for which $l(u, v) < l(v, w)$.

The following lemma says that when looking for good algorithms, it is enough to consider routing algorithms derived from a labeling.

Lemma 1. *The following are equivalent for a routing algorithm:*

1. *There are finitely many s, r-routings.*
2. *One of its strengthenings is derived from a labeling.*
3. *There is no essential cycle.*

Proof. ($1 \Rightarrow 2$) Given a routing algorithm with only finitely many s, r-routings, for every edge let $l(u, v)$ be the maximum i such that there is an s, r-routing $v_0, \ldots, v_k$ such that $v_{i-1} = u$ and $v_i = v$. Let $l(u, v) = 0$ if there is no such routing, that is, if uv is not in any essential instruction. Note that for any essential instruction uvw, there must be a u, r-routing $u, v, w, \ldots, r$. Letting $v_0, \ldots, v_k$ be an s, r-routing that achieves $l(u, v)$, the routing $v_0, \ldots, v_{i-1}(= u), v_i(= v), w, \ldots, r$ shows that $l(u, v) < l(v, w)$. Thus the routing algorithm derived from l is a strengthening of the given algorithm.

($2 \Rightarrow 3$) Let l be a labeling. It suffices to show that the routing algorithm derived from l has no essential cycle $v_0, \ldots, v_k$. This is immediate since otherwise $l(v_0 v_1) < \ldots < l(v_k v_{k+1}) = l(v_0 v_1)$.

($3 \Rightarrow 1$) Infinitely many s, r-routings in a finite graph mean that there must exist s, r-routings of arbitrary length. However, in a routing $v_0, \ldots, v_k$ with $k > 2|E(G)|$, an edge must appear twice in the same direction. Then there is an essential cycle. $\qquad\qquad\square$

We remark that the equivalence between the first and third properties was essentially already presented in Section 7 of [5]. Also, note that it follows from the proof that for any routing algorithm derived from a labeling, not just are there only finitely many messages received at r, but in fact there are only finitely many messages generated overall.

Finally, we say that a routing algorithm is *k-robust* if (i) there are only finitely many s, r-routings and (ii) whenever at most k edges are removed such that some s, r-path remains in G, then an s, r-routing remains as well. The *robustness* of a graph G, denoted by $\rho(G)$, is the maximum robustness of any of its routing algorithms. If G has a k-robust algorithm for all k, then we say that G is *infinitely robust*. It is shown in [5] that a graph is infinitely robust if and only if there is no essential cycle in the routing algorithm formed by taking every instruction that appears along some s, r-path.

3   A General Lower Bound: Beating Menger's Theorem

Menger's Theorem is a fundamental result in graph connectivity. It states that the maximum size of a set of edge-disjoint s, r-paths in G equals the minimum size of an edge-set whose removal separates s and r (denoted $\kappa'_G(s, r)$, following [6]). Since routing along k edge-disjoint s, r-paths is a $(k - 1)$-robust algorithm, this immediately yields $\rho(G) \geq \kappa'_G(s, r) - 1$. In fact, Menger's Theorem can be interpreted as a sharp routing result for the simpler model where an algorithm must always route a message from s to r whenever at most k edges are removed. Since a k-robust algorithm is not required to route a message successfully when the removed edges separate s from r, there is the potential for a stronger result. We prove such a result.

Theorem 1. *For any connected graph G with distinct vertices s and r, $\rho(G) \geq \kappa'_G(s, r)$.*

Theorem 1 yields a constant lower bound on robustness for all graphs.

Corollary 1. *For any connected graph G with distinct vertices s and r, $\rho(G) \geq 2$.*

Proof. If G has no cut-edge that separates s and r, then $\kappa'_G(s, r) \geq 2$, so $\rho(G) \geq 2$. Otherwise let uv be such an edge, let G_1 be the graph formed from the component of $G - uv$ that contains s and u, treating u as the new r, and similarly define G_2 with r, v. By induction, G_1 and G_2 have 2-robust algorithms A_1 and A_2 derived from labelings of G_1 and G_2. By shifting each labeling by a constant, we may assume that every label is positive, and that for some number q, every label in G_1 is less than q and every label in G_2 is greater than q. Let $l(u, v) = q$ and let $l(v, u) = 0$. Let A' be the algorithm derived from l. We claim that A' is a 2-robust algorithm for G. By Lemma 1, there are only finitely many s, r-routings. Let X be any set of at most 2 edges of G whose removal leaves an s, r-path P in G. Then u, v is a subpath of P, and the s, u-portion and v, r-portion of P are s, r-paths in G_1 and G_2, respectively. Therefore A_1 and A_2 contain s, r-routings in $G_1 - X$ and $G_2 - X$, respectively. Since these routings use labels less than and greater than q, respectively, together with uv they form an s, r-routing in $G - X$. □

The following examples show that these results are sharp.

Example 1. Let F_0 be the graph obtained from $K_{3,3}$ by letting the partite sets be $\{s, u, v\}$ and $\{r, w, x\}$ and deleting the edge sr. (F_0 is part of the forbidden minor characterization of infinitely robust graphs in [5].) It is easy to see that $\rho(F_0) \leq 2$: If A were a 3-robust algorithm, then removing $\{sx, vw, ur\}$ and $\{sw, xu, vr\}$ forces s, w, u, x, v, r and s, x, v, w, u, r to be routings, respectively, but then w, u, x, v, w is an essential cycle. For $k \geq 2$, let G_k be the graph created by adding $k - 2$ internally disjoint s, r-paths to F_0. One can easily modify the above argument to show that $\rho(G_k) \leq 2 + (k - 2)$, and since $\kappa'_{G_k}(s, r) = k$, $\rho(G_k) = k$. In fact it can be seen that identifying F_0 with any graph G for which $\kappa'_G(s, r) = k - 2$ at s and r yields a graph G' such that $\kappa'_{G'}(s, r) = k = \rho(G')$.

Remark 1. It is easy to see that robustness is unchanged by subdividing edges. Since infinitely robust graphs have a forbidden minor characterization [5], it is natural to ask whether there are similar characterizations for finite robustness. Let H_k be the graph obtained from G_k and F_0 by identifying their vertices labeled s, adding an edge e joining their vertices labeled r, and letting the special vertex r for G'_k be the r in F_0. Using Example 1 it is not hard to see that $\rho(H_k) = 3$. Also, deleting e from H_k lowers the robustness to 2, and contracting e raises the robustness to $k + 2$. It is not hard to check that for any edge e' in F_0, $F_0 - e'$ is infinitely robust (see also [5]); it follows that $H_k - \{e, e'\}$ is infinitely robust, and one can show that the robustness of $H_k - e'$ is at least $k + 1$. Thus, a minor characterization seems rather unlikely.

We only give a descriptive outline of the proof of Theorem 1 because the details are somewhat technical. The complete proof will be given in the final version of the paper. The proof goes more smoothly by allowing graphs to have multiple edges. For technical reasons, for each routing $s, e, \ldots, r$ of a routing algorithm, we consider $\emptyset se$ to be an instruction.

We begin with a very important special case.

Lemma 2. *Let G be a graph such that its s, r-edge-cuts of size $\kappa'_G(s, r)$ are precisely $\{sv: sv \in E(G)\}$ and $\{vr: vr \in E(G)\}$. Then $\rho(G) \geq \kappa'_G(s, r)$.*

Proof (sketch). Let $k = \kappa'_G(s, r)$. By Menger's Theorem, we may let $\mathcal{P} = \{P_1, \ldots, P_k\}$ be a set of edge-disjoint s, r-paths with minimum total number of edges. It is easy to check that minimum size ensures that when two paths of $\mathcal{P}$ intersect in two vertices other than s and r, the vertices will appear in the same order in both paths.

Given an algorithm A, an A-*cut* is a set of k edges such that removing them destroys all s, r-routings and does not separate s from r in G. Note that if A' is a strengthening of A, then each A'-cut is an A-cut, so the set of A-cuts contains the set of A'-cuts.

Our initial algorithm, A_0, will be the set of instructions that sends a packet from s along each path of $\mathcal{P}$. Then A_0 is $k - 1$-robust; we strengthen it to a k-robust algorithm by defining a sequence of algorithms $A_1, A_2, \ldots$, such that each subsequent algorithm A_i will contain all instructions of previous algorithms. Thus, any A_i-cut will consist of one edge from each path of $\mathcal{P}$. Let G_i be the graph induced by all the instructions in A_i.

We can partition the instructions in A_i as follows: Consider the graph where the vertices are the instructions of A_i and $e_1 v_1 f_1$ is adjacent to $e_2 v_2 f_2$ if and only if $v_1 \neq v_2$ and either $f_1 = e_2$ or $f_2 = e_1$. Partition A_i into $A_i^1, \ldots, A_i^{k_i}$ such that each A_i^j is the vertex set of a component of the auxiliary graph. For each $1 \leq j \leq k_i$, let G_i^j be the subgraph of G_i induced by the instructions in A_i^j, and let $\mathcal{P}_i^j \subseteq \mathcal{P}$ be the set of paths contained in G_i^j. Note that $k_0 = k$, and for each $1 \leq j \leq k_0$, $G_0^j = P_j$ and $\mathcal{P}_0^j = \{P_j\}$. One can show that for each A_i, the edges of G_i are partitioned by $G_i^1, \ldots, G_i^{k_i}$ and $\mathcal{P}_i^1, \ldots, \mathcal{P}_i^{k_i}$ partitions $\mathcal{P}$.

For each A_i we also designate a special vertex y_i^P in $P - s$ for each $P \in \mathcal{P}$.

Let z_i^P be the vertex after y_i^P on P. Each algorithm A_i will have the following important properties:

1. A_i has no essential cycle.
2. If X is a set of edges of G that contains exactly one edge from each path in $\mathcal{P}$ and whose deletion does not separate s from r, then X is an A_i-cut if and only if:
   (a) for each $P \in \mathcal{P}$, no edge of the s, y_i^P-portion of P is contained in X except possibly the first edge (which is incident to s), and
   (b) for each $1 \leq j \leq k_i$, either X contains the first edge of every path in $\mathcal{P}_i^j$ or X doesn't contain the first edge of any path in $\mathcal{P}_i^j$.
3. For each $1 \leq j \leq k_i$, A_i^j is the union of the instructions along certain s, r-paths, where each path begins with the first edge of a path in $\mathcal{P}_i^j$ and ends with the y_i^P, r-portion of some $P \in \mathcal{P}_i^j$.

Observe that these properties hold true for A_0 once we let y_0^P be the second vertex on P for each $P \subset \mathcal{P}$.

Procedure to obtain A_{i+1} from A_i:

For fixed i and for each $1 \leq j \leq k_i$, let X_i^j be the edge set consisting of the first edge of each path in $\mathcal{P} - \mathcal{P}_i^j$ and the edge $y_i^P z_i^P$ for each $P \in \mathcal{P}_i^j$. Note that X_i^j does not separate s from r in G by the statement of this lemma, and hence by property 2, X_i^j is an A_i-cut. Let H_i^j be obtained from G_i^j by deleting the edges of the y_i^P, r-subpath of each $P \in \mathcal{P}_i^j$ and letting H_i^j be the component that contains s. (See Fig. 1 for an example.) Since $G - X_i^j$ does not separate s from r in G, we may let Q_i^j be a minimal path in $G - X_i^j$ from H_i^j to the graph formed by taking the union of $H_i^{j'} - s$ for all $j' \neq j$ and the z_i^P, r-portion of each path $P \in \mathcal{P}$.

Case 1. For some distinct j, j', Q_i^j ends in $H_i^{j'}$.

In short, we add instructions that pass a message through Q_i^j in both directions, which defines A_{i+1} such that $k_{i+1} = k_i - 1$, where A_i^j and $A_i^{j'}$ are replaced by the single set $A_i^j \cup A_i^{j'} \cup (A_{i+1} - A_i)$, G_i^j and $G_i^{j'}$ are replaced by $G_i^j \cup G_i^{j'} \cup Q_i^j$, $\mathcal{P}_i^j$ and $\mathcal{P}_i^{j'}$ are replaced by their union, and not much else is changed.

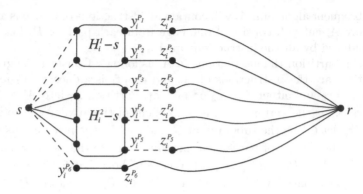

Fig. 1. $G_i - X_i^2$, where $\kappa(s,r) = 6$, $k_i = 3$, and $V(H_i^3) = \{s, y_i^{P_6}\}$

Case 2. For each $1 \leq j \leq k_i$, Q_i^j ends in the z_i^P, r-portion of some path $P \in \mathcal{P}$. If we think of this as defining a digraph on $[k_i]$ where j is adjacent to $j' \in [k_i]$ if Q_i^j ends on $P \in \mathcal{P}_i^{j'}$, then this digraph has a cycle $j_1, \ldots, j_m, j_1$ for some $m \geq 1$. (Treat the indices modulo m.) Note that by minimality, each pair of such paths can only intersect at their last vertex, which cannot be r by the statement of this lemma.

For $1 \leq l \leq m$, we add instructions that send a message from H_i^j along Q_i^j and into a path in $\mathcal{P}_i^{j_{l+1}}$. This defines A_{i+1} such that $k_{i+1} = k_i - m + 1$, joining parts $j_1, \ldots, j_m$ from the partitions of the A_i, G_i, and $\mathcal{P}$. Also, the endpoint of $Q_i^{j_l}$ on the z_i^P, r-portion of some $P \in \mathcal{P}_i^{j_{l+1}}$ is assigned to be y_{i+1}^P. Not much else changes.

This process ends when $k_i = 1$ and $\{y_i^P z_i^P : P \in \mathcal{P}\}$ is the set of edges adjacent to r. Then there are no A_i-cuts, so A_i is k-robust. □

For the proof of Theorem 1, we again let $\mathcal{P}$ be a set of $\kappa_G'(s,r)$ edge-disjoint s, r-paths in a given graph G. Let $\mathcal{S}$ be the union of all edge sets of size $\kappa_G'(s,r)$ whose deletion separates s from r. Then each edge in $\mathcal{S}$ lies on some path in $\mathcal{P}$. Note that each component of $G - \mathcal{S}$ contains an endpoint of an edge in $\mathcal{S}$, so each contains a vertex on a path in $\mathcal{P}$. Also, together the components of $G - \mathcal{S}$ contain all of $V(G)$. For each component C of $G - \mathcal{S}$, let $\mathcal{P}_C$ be the set of paths in $\mathcal{P}$ that intersect C. Let G_C be the graph formed from the union of C, the edges immediately preceding C on each path of $\mathcal{P}$, with the endpoints not in C identified to a single vertex labeled s, and the edges immediately following C on each path of $\mathcal{P}$, with the endpoints not in C identified and labeled r. This could create multiple edges, which is why it is convenient to give our proofs for multigraphs.

We first assume that s and r form single-vertex components of $G - \mathcal{S}$, and for every other component C we apply Lemma 2 to G_C, then join together the resulting algorithms to produce an algorithm for G as desired. The proof requires several short lemmas about the related structure of $\mathcal{P}$, $\mathcal{S}$, and the components of $G - \mathcal{S}$. Afterward, it is not hard to extend the proof to the general case.

4   Some Special Families of Graphs

We can frequently do better than the lower bound $\kappa'(s,r)$. As a first example, consider the complete graph K_n on n vertices with two distinct vertices labeled s and r. It is easy to see that K_n is infinitely robust for $n \leq 4$ (see also [5]). Note that for this graph, $\kappa'(s,r) = n - 1$.

Theorem 2. *If $n \geq 5$, $\rho(K_n) = 2n - 5$.*

Proof. Consider the routing algorithm that is derived from $l(s,v) = 1$ and $l(v,s) = 0$ for all $v \neq s$, $l(u,v) = 2$ for all distinct $u, v \notin \{s,r\}$, and $l(v,r) = 3$ and $l(r,v) = 0$ for all $v \notin \{s,r\}$. By Lemma 1 this algorithm has only finitely many s,r-routings. Let X be a set of edges whose removal destroys all s,r-routings but does not disconnect s from r in K_n. Let k and k' be the number of edges of the form sv and vr in $K_n - X$, respectively. Since all paths of the form s, v, r are routings, those edges have distinct endpoints in $V(K_n) - \{s,r\}$, and since all paths of the form s, u, v, r are routings, X must contain at least kk' edges uv such that $u, v \notin \{s, r\}$. Also, s, r is a routing so $sr \in X$, and $|X| \geq 1 + (n - 2 - k) + (n - 2 - k') + kk' = 2n - 4 + (k - 1)(k' - 1) \geq 2n - 4$. Therefore this algorithm is $(2n - 5)$-robust.

To see that this is optimal, suppose that A is a $(2n - 4)$-robust algorithm. Consider any distinct vertices $u, v \notin \{s, r\}$. After removing the edges $\{sw: w \neq u\} \cup \{uw: w \notin \{s, v\}\} \cup \{vr\}$, a path s, u, v, w, r remains for any vertex $w \notin \{s, u, v, r\}$, so some s, r-routing must also remain. Its initial vertices must be s, u, v, w for some vertex $w \notin \{s, u, v, r\}$, so there must be an essential instruction uvw for some $w \notin \{s, u, v, r\}$. Since u and v are arbitrary vertices of $V(K_n) - \{s, r\}$, this argument also implies an essential instruction of the form vww' for some w', and so forth. This generates a walk of arbitrary length on $V(K_n) - \{s, r\}$, and eventually two consecutive vertices must repeat and we obtain an essential cycle. A cannot have an essential cycle, and this contradiction completes the proof. $\square$

Remark 2. In general, in order to show that a graph G is not k-robust, we may give various sets of at most k edges whose deletion does not separate s from r in the graph. For each such set X, we say that an instruction which is on every s, r-path in $G - X$ is *forced*, because some s, r-routing in a k-robust algorithm must contain that instruction. If we produce a set of forced instructions that form an essential cycle, then that implies that there can be no k-robust algorithm for G and $\rho(G) < k$.

Next we consider a type of modified grid ("mesh" in [2]). Let $G(m, n)$ be the graph with vertex set $\{v_{ij}: 1 \leq i \leq m, 1 \leq j \leq n\} \cup \{s, r\}$ such that two vertices other than s, r are adjacent exactly when one index is the same and the other differs by 1, $N(s) = \{v_{i1}: 1 \leq i \leq m\}$, and $N(r) = \{v_{in}: 1 \leq i \leq m\}$. (See Fig. 2.) That is, $G(m, n)$ is the product of paths of m and n vertices, with a sender and receiver connected to opposite ends. It is easy to see that if $n = 1$ or $m = 1, 2$ then $G(m, n)$ is infinitely robust (see also [5]). Note that for this graph, $\kappa'(s, r) = m$.

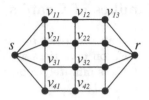

Fig. 2. $G(4,3)$

Theorem 3. *If $n \geq 2$ and $m \geq 3$, then $\rho(G(m,n)) = m + 2$.*

Proof. We show that $G(m,n)$ is not $(m+3)$-robust by finding sets of $m+3$ edges that force the instructions that form the essential cycle $v_{21}, \ldots, v_{m1}, v_{m2}, v_{(m-1)2}$, $v_{(m-1)1}, \ldots, v_{11}, v_{12}, v_{22}, v_{21}$. In particular, if we remove $sv_{21}, \ldots, sv_{m1}, v_{11}v_{21}$, $v_{12}v_{13}$, $v_{22}v_{23}$, and $v_{22}v_{32}$ (if $n = 2$ then $v_{i3} = r$), then at least one s, r-path remains, and all remaining s, r-paths begin with $s, v_{11}, v_{12}, v_{22}, v_{21}, v_{31}$. Also, for each $1 < i < m$, if we remove $\{sv_{j1} : j \neq i-1\}$, $v_{(i-2)1}v_{(i-1)1}$, $v_{(i-1)1}v_{(i-1)2}$, and $v_{i1}v_{i2}$, then $v_{(i-1)1}v_{i1}v_{(i+1)1}$ is forced. By symmetry, every desired instruction can be forced, so according to Remark 2, $\rho(G(m,n)) \leq m + 2$.

Consider the algorithm derived from the following labeling: $l(s, v_{i1}) = 1$, $l(v_{ij}, v_{i(j+1)}) = mj + 1$, and $l(v_{in}, r) = mn + 1$ for $1 \leq i \leq m$ and $1 \leq j < n$; $l(v_{(i-1)j}, v_{ij}) = m(j-1)+i$ and $l(v_{ij}, v_{(i-1)j}) = m(j-1)+m-i+2$ for $2 \leq i \leq m$ and $1 \leq j < n$; every other label is 0. Note that the s, r-routings are precisely the s, r-paths (as walks) which don't go "backward": that is, v_{ij} can precede v_{kl} in the routing only if $j \leq l$. We claim that this algorithm is $(m + 2)$-robust. By Lemma 1, it contains only finitely many s, r-routings.

Let X be a set of edges such that deleting X prevents the algorithm from sending a message from s to r and $G - X$ still contains an s, r-path. Let P be a shortest s, r-path in $G - X$. Then P must contain a "backward" subpath for which the second index of the vertices is decreasing; let $v_{ik}, \ldots, v_{ij}$ with $j < k$ be the last maximal backward subpath on P. By the choice of P, X contains either an edge $v_{il}v_{i(l+1)}$ for some $l \geq k$ or the edge $v_{in}r$. Without loss of generality, we can assume that v_{ij} is followed by $v_{(i+1)j}$ in P. By the choice of v_{ij}, the second index on vertices along the v_{ij}, r-subpath of P is nondecreasing. Hence that subpath contains a vertex v_{lk} for some $l > i$; by choice of P, X must contain an edge of the path $v_{ik}, \ldots, v_{lk}$. Also by choice of P, X contains either an edge $v_{i(l-1)}v_{il}$ with $l \leq j$ or the edge sv_{i1}. For topological reasons the s, v_{ik}-subpath of P must intersect the path $v_{1j}, \ldots, v_{(i-1)j}$. Then by the choice of P, X contains $v_{(l-1)j}v_{lj}$ for some $l \leq i$. That makes four edges in X so far, none of which are on the paths $s, v_{l1}, \ldots, v_{ln}, r$ for $l \neq i$. Since X must also contain at least one edge from each of those paths, $|X| \geq (m-1) + 4$ and the algorithm is $(m + 2)$-robust. □

Another interesting communication network is the m-dimensional hypercube, or m-*cube*. We represent the vertices as $0, 1$-vectors of length m. The *weight* of a vertex is its number of 1s. Let s and r be the unique vectors of weight 0 and m, respectively. Two vertices are adjacent if they differ in exactly one coordinate. We denote the m-cube by Q_m.

Clearly Q_m is infinitely robust for $m = 1, 2$. Note that $\kappa'(s, r) = m$.

Theorem 4. *If $m \geq 3$, then $\rho(Q_m) = 3m - 6$.*

Proof. Consider any distinct i, j between 0 and m. Then let u be the vertex with a 1 in coordinate i and 0s elsewhere and let v be the vertex with 1s in coordinates i and j and 0s elsewhere. If we delete the edges from s to each weight 1 vertex other than u, the edges from u to each weight 2 vertex other than v, and each edge from v to a weight 3 vertex, then every remaining s, r-path begins s, u, v, x, y for some weight 1 vertex $x \neq u$ and some weight 2 vertex $y \neq v$. Since $i.j$ are chosen arbitrarily we can repeat the argument with x, y instead of u, v. Thus deleting sets of $3m - 5$ edges will force an infinite walk of essential instructions among the vertices of weights 1 and 2. This implies an essential cycle and by Remark 2, $\rho(Q_m) \leq 3m - 6$.

A $(3m-6)$-robust algorithm will be presented in the full version of the paper.

$\square$

Note that the optimal algorithm given for a hypercube could be described as a "vertex-label algorithm": we assign labels to the vertices, i.e., their weights, and only send messages from vertices of lower labels to vertices of higher labels. The optimal algorithm for the grid $G(m, n)$ seems to have something similar if we partition its vertices into n "levels" consisting of the vertices at a fixed distance from s, and label the vertices in a level according to that distance. There are additional instructions needed within a level and to enter or exit a level, but they are relatively simple. The complete graph even has something like this if we think of all its non-root vertices as constituting a single level. All algorithms thus far have some kind of level-structure for which the algorithm primarily consists of instructions that send a message forward through the levels.

The next example shows that in general the situation can be more complicated, as any optimal algorithm for it cannot be separated into linearly-ordered levels on which the algorithm is "easy" to describe. Thus the levels in the previous examples seem to be more an artifact of those particular examples, rather than part of a general truth. Furthermore, the optimal algorithms for this example are surprisingly asymmetrical, considering the symmetry of the graph itself.

Let M_8 be the graph shown in Fig. 3, with vertex set $\{s, r, 1, 2, 3, 4, 5, 6\}$ and edge set $\{s1, 12, 23, 3r, s4, 45, 56, 6r, 16, 25, 34\}$. This is a rooting of the Möbius ladder of order 8 minus the edge sr. In general, adding an edge from s to r changes robustness by exactly 1, so one can study the smaller graph. Clearly $\kappa'(s, r) = 2$, and note that the graph is symmetric on switching $1, 2, 3$ with $4, 5, 6$.

Theorem 5. *$\rho(M_8) = 3$, and no 3-robust algorithm is symmetric on switching $1, 2, 3$ with $4, 5, 6$.*

Proof. Deleting edge sets $\{s4, 16, 25, 3r\}$ and $\{s1, 34, 25, 6r\}$ forces instructions that form the essential cycle $1, 2, 3, 4, 5, 6, 1$. Hence $\rho(M_8) \leq 3$.

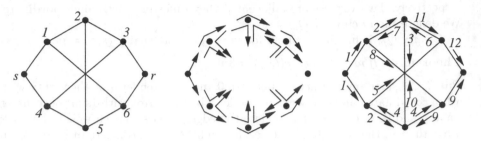

Fig. 3. M_8, instructions in any 3-robust routing algorithm, a labeling

We will show that deleting edge sets of size 3 necessitate certain instructions, and that while they can be completed to form a 3-robust algorithm, it cannot be done symmetrically with respect to switching 1,2,3 and 4,5,6. Deleting $\{s4, 16, 25\}$ forces instructions $s12$ and 123, deleting $\{16, 34, 56\}$ forces $23r$, deleting $\{12, 56, 3r\}$ forces $s16$ and $16r$, deleting $\{s4, 16, 23\}$ forces 125, and deleting $\{12, 34, 6r\}$ forces 523. Also, deleting $\{s1, 45, 3r\}$ forces 432 and deleting $\{s1, 56, 3r\}$ forces 216. By symmetry we could make the same arguments but with $1, 2, 3$ switched with $4, 5, 6$, which forces a symmetric set of instructions. (See Fig. 3 for the set of instructions forced so far.) It is easy to see that so far there is no essential cycle.

Note that $\{s1, 45, 3r\}$ forces either 321 or 325, and $\{s1, 56, 3r\}$ forces either 321 or 521. Symmetry implies that the instruction set also contains 654 or $\{652, 254\}$. Adding 321 and 654 creates the essential cycle $1, 6, 5, 4, 3, 2, 1$, and adding $\{325, 521, 652, 254\}$ creates a few essential cycles including $1, 6, 5, 2, 1$, $4, 3, 2, 5, 4$, and $1, 6, 5, 2, 3, 4, 5, 2, 1$. However, adding $\{321, 652, 254\}$ (or by symmetry, $\{654, 325, 521\}$) will create a 3-robust algorithm. This can be checked without too much case analysis: First, verify that this routing algorithm is basically identical to the algorithm derived from the labeling in Fig. 3. Hence there is no essential cycle. Any set of 3 edges that might defeat the algorithm must intersect the s, r-routings $s123r$, $s456r$, $s16r$, and $s43r$, so must include at least one edge from $\{s1, s4, 3r, 6r\}$, as well as at most one edge from each set $\{s1, s4\}$ and $\{3r, 6r\}$. This helps to organize the necessary case analysis which will complete the proof. $\square$

5   Open Problems

Since this paper is the first to discuss robustness as such, there are many questions that can be posed. We mention two.

The proof of Theorem 1 actually constructs a $\kappa'(s, r)$-robust algorithm in polynomial time. This leads to the general question of how k-robust algorithms can be generated quickly—for $k = \rho(G)$, for k as a different function of s, r-edge-connectivity, or for k chosen in some other way. For that matter, how quickly can the robustness of a given graph be computed? When Lovász [4] suggested the study of finite robustness, he also suggested another possibility, namely, modifying the model such that edges fail according to some probability p. This question remains wide open.

References

1. Adler, M., Fich, F.E.: The Complexity of End-to-End Communication in Memory-less Networks. 18th Annual ACM Symposium on Principles of Distributed Computing, May 1999, 239–248.
2. Adler, M., Fich, F.E., Goldberg, L.A., Paterson, M.: Tight Size Bounds for Packet Headers in Narrow Meshes. Proceedings of the Twenty Seventh International Colloquium on Automata, Languages and Programming (ICALP), July 2000.
3. Fich, F.E.: End-to-end Communication. Proceedings of the 2nd International Conference on Principles of Distributed Systems, Amiens, France, (1998), 37–43.
4. L. Lovász: Personal communication, 2003.
5. Fich, F.E., Kündgen, A., Pelsmajer, M.J., Ramamurthi, R.: Graph Minors and Reliable Single-Message Transmission, accepted for publication in SIAM J. Discr. Math.
6. West, D.B.: Introduction to Graph Theory (Second Edition). Prentice-Hall, Inc., Upper Saddle River, NJ, 2001.

Stable Local Scheduling Algorithms With Low Complexity and Without Speedup for a Network of Input-Buffered Switches

Claus Bauer

Dolby Laboratories,
San Francisco, CA, 94103, USA
cb@dolby.com

Abstract. The choice of the scheduling algorithm is a major design criteria for internet switches and routers. Research on scheduling algorithms has mainly focused on maximum weight matching scheduling algorithms, which are computationally very complex and on the computationally less complex maximal weight matching algorithms which require a speedup of two to guarantee the stability of the switch. For practical purposes, neither a high computational complexity nor a speedup is desirable. In this paper, we propose a specific maximal weight matching algorithm that guarantees the stability of a single switch without a speedup.

Whereas initial research has only focused on scheduling algorithms that guarantee the stability of a single switch, it is known that scheduling algorithms that guarantee the stability of individual switches do not necessarily stabilize networks of switches. Recent work has shown how scheduling algorithms for single switches can be modified in order to design scheduling algorithms that stabilize networks of input-queued switches. We apply those results to the design of the maximal weight matching algorithm proposed in this paper and show that the algorithm does not only stabilize a single switch, but also networks of input-queued switches.

1 Introduction

The introduction of new fast optical transmission technologies such as Dense Wavelength Division Multiplexing (DWDM) has dramatically increased the transmission capacity of optical fibers. In consequence, there is a need for switches and routers that work at or above the speed of the high-speed optical links connecting them. As purely output-buffered switches have shown to be impractical due to the required high speedup in the switching core, in this paper, we only consider input-buffered (IQ) or combined input/output (CIOQ) buffered switches. A typical architecture of CIOQ switch is given in figure 1. For each input i, there are N virtual output queues $VOQ_{i,j}$, $1 \leq j \leq N$. The cells arriving at input i and destined for output j are buffered in $VOQ_{i,j}$. The switching core itself is modeled as a crossbar which requires that not more than one packet can be

A. López-Ortiz and A. Hamel (Eds.): CAAN 2004, LNCS 3405, pp. 102–113, 2005.

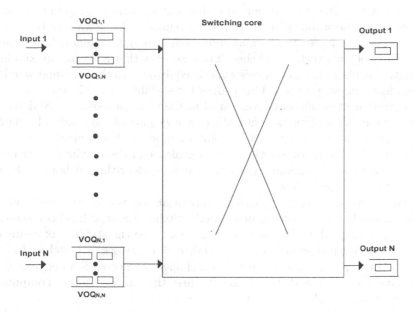

Fig. 1. Architecture of an input queued switch

sent simultaneously from the same input or to the same output. Many switch architectures require the switch to work at a speedup of S, $S \geq 1$, i.e., the switch works at a speed S times faster than the speed of the input/output links.

The design of the scheduling algorithm of a switch determines the loss and delay characteristics of a switch. Its computational complexity influences the frequency at which the switch configuration can be updated. In [6] and [7], it has been shown that for a speedup of $S = 1$, maximum weight matching algorithms with specific weights provide guarantees on the throughput of a switch and the average delay experienced by a packet arriving at the switch.

Because maximum weight matching algorithms are difficult to implement due to their computational complexity of $O(N^3)$, researchers have investigated less complex scheduling algorithms that can stabilize CIOQ switches. In [5] and [4], it has been shown that maximal weight matching algorithms that are deployed with a speedup of two or even slightly less can guarantee the stability of an individual CIOQ switch. The computational complexity of a maximal weight matching algorithm is $O(N^2)$. A definition of maximal weight matching algorithms is given in section 4.1.

Due to the impracticality of even low speedups at high line speeds, it is desirable to understand if maximal weight matching algorithms that are stable with a speedup of $S = 1$ exist. In [8], it has been shown that under the assumption that no two $VOQs$ ever have the same length, such a maximal weight matching algorithm exists. In practice, this assumption can obviously not be made.

This paper proposes the first maximal weight matching algorithm that under reasonable assumptions on the traffic arrival patterns, as defined in section 2.2, stabilizes a CIOQ switch without any speedup and without assumptions on the $VOQs$. We prove the stability of the proposed algorithm using an idea from [8]: We first show that a large class of maximum weight matching algorithms guarantees stability of a network of switches. Then we show that a specific maximum weight matching algorithm out of this class is equivalent to the maximal weight matching algorithm we propose. This proves the stability of the latter.

The improvement of our work compared to the ideas proposed in [8] derives from the specific choice of our weights. We define weights of the maximal weight matching algorithm such that at any point in time, no two weights are ever equal. This fact is crucial for showing the equivalence of the maximal matching algorithm with a specific maximum weight matching algorithm without making any assumptions on the $VOQs$.

The early work on switch algorithms quoted above has investigated stable scheduling algorithms for single input-queued switches. Later, it has been shown in [2] and [3] that scheduling algorithms that guarantee the stability of a single switch do not necessarily guarantee the stability of a network of switches. In [2] and [3], switching policies that require the exchange of information between the switches have been proposed. In [1], for the first time a distributed, computationally complex maximum weight matching algorithm that does not require the exchange of information between switches in the network has been proposed. In this paper, we apply the methods developed in [1] to the design of the maximal weight matching algorithm proposed here. In consequence, our algorithm does not only guarantee the stability of a single switch, but also stabilizes a networks of CIOQ switches when it is deployed without a speedup at all switches in the network.

In conclusion, the algorithm proposed in this paper satisfies *the most common performance requirements on a scheduling algorithm:* It guarantees the stability of a network of switches, it can be implemented in distributed manner that does not require the exchange of information between the switches, it is of feasible computational complexity and it does not require any speedup.

This paper is organized as follows. In the next section, we introduce the terminology to describe a network of switches. In section 3, we prove stability results for a large class of maximum weight matching algorithms. In section 4, we design a maximal weight matching algorithm and show that it stabilizes networks of input-queued switches. We conclude in section 5.

2   Terminology and Model

2.1   Model of a Network of Queues

In this section, we follow an approach in [1] to describe our model of a queueing system. We assume a system of J physical queues $\tilde{q}^j$, $1 \leq j \leq J$ of infinite capacity. Each physical queue consists of one or more logical queues, where each logical queue corresponds to a certain class of customers within the physical queue. Whenever a packet moves from one physical queue to another, it changes class and therefore also changes logical queue. We denote a logical queue by q^k, $1 \leq k \leq K$, where $K \geq J$. A packet enters the network via an edge switch, travels through a number of switches and leaves the network via another edge switch. We define a function $L(k) = j$ that defines the physical queue $\tilde{q}^j$ at which packets belonging to the logical queue q^k are buffered. The inverse function $L^{-1}(j)$ returns the logical queues q^k that belong to the physical queue $\tilde{q}^j$.

Throughout this paper, the time t is described via a discrete, slotted time model. Packets are supposed to be of fixed size and a timeslot is the time needed by a packet to arrive completely at an input link.

We define a row vector $X_n = (x_n^1, ..., x_n^K)$, where the k-th vector x_n^k represents the number of packets buffered in the logical queue q^k in the n-th timeslot. We define $E_n = (e_n^1, ..., e_n^K)$, where e_n^k equals the number of arrivals at the logical queue q^k in the n-th timeslot. Analogously, we define $D_n = (d_n^1, ..., d_n^K)$, where d_n^k expresses the number of departed packets from q^k in the n-th timeslot. Thus, we can describe the dynamics of the system as follows:

$$X_{n+1} = X_n + E_n - D_n. \tag{1}$$

Packets that arrive at a logical queue q^k either arrive from outside the system or are forwarded from a queue within the system. Therefore,

$$E_n = A_n + T_n,$$

where $A_n = (a_n^1, ..., a_n^K)$ denotes the arrivals from outside the system and $T_n = (t_n^1, ..., t_n^K)$ denotes the arrivals from inside the system.

We further define a routing matrix $R = [r_{i,j}]$, $1 \leq i, j \leq K$, where $r_{i,j}$ is the fraction of customers that depart from the logical queue q^i and are destined for the logical queue q^j. Assuming a deterministic routing policy, there holds, $r_{i,j} \in \{0, 1\}$, $\sum_{1 \leq i \leq K} r_{i,j} \leq 1$, $\sum_{1 \leq j \leq K} r_{i,j} \leq 1$. We set $r_{i,j} \neq 0$, if q^j follows q^i along the route. Noting that $T_n = D_n R$ and writing I for the identity diagonal matrix, we find

$$X_{n+1} = X_n + A_n - D_n(I - R). \tag{2}$$

We assume that the external arrival processes are stationary and satisfy the Strong Law of Large Numbers. Thus,

$$\lim_{n \to \infty} \frac{\sum_{i=1}^{n} A_i}{n} = \Lambda \qquad \text{w.p.1}, \tag{3}$$

where $E[A_n] = \Lambda = (\lambda_1, .., \lambda_K), \forall n \geq 1$[1]. Noting that $(I-R)^{-1} = I + R + R^2 + ...$, we find that the average workload $W = (w^1, ..., w^K)$ at the logical queues q^k is given by $W = \Lambda(I - R)^{-1}$.

Finally, we recall a stability criteria for a network of queues from [1].

---

[1] Throughout the paper, we abbreviate "with probability 1" by "w.p.1".

Definition 1. *A system of queues is rate stable if*

$$\lim_{n \longrightarrow \infty} \frac{X_n}{n} = \lim_{n \longrightarrow \infty} \frac{1}{n} \sum_{i=0}^{n-1} (E_i - D_i) = 0 \qquad w.p.1.$$

A necessary condition for the rate stability of a system of queues is that the average number of packets that arrive at any physical queue $\tilde{q}^j$ during a timeslot is less than 1. In order to formalize this criteria, we introduce the following definition:

Definition 2. *For a vector $Z \in \mathbb{R}^K$, $Z = (z^1, .., z^K)$, and the function $L^{-1}(k)$ as defined in this subsection, we set:*

$$\|Z\|_{maxL} = \max_{j=1,..,J} \left\{ \sum_{k \in L^{-1}(j)} z^k \right\}. \tag{4}$$

The necessary condition for rate stability can be formalized as follows:

$$\|W\|_{maxL} < 1. \tag{5}$$

In the sequel, we will say that a system of networks that satisfies condition (5) and is rate stable achieves 100% throughput.

2.2   Model of a Network of Switches

In this section, we apply the terminology of the previous section to a network of IQ/CIOQ switches. A network of IQ/CIOQ switches can be conceived as a queueing system as defined in the previous section where the virtual output queues are considered as the physical queues. In this model we neglect the output queues of the switches because instability can only occur at the VOQs (see [1]).

We say that packets that enter the network via the input of a given switch and leave the network via the output of a given switch belong to the same flow. Packets belonging to the same flow travel through the same sequence of physical queues and are mapped to the same logical queues at each physical queue, i.e., a flow can be mapped biunivocally (or one-to-one) to a series of logical queues.

We assume a *per-flow* FIFO scheduling scheme. It is shown in [1] how stable *per-flow* scheduling schemes can be used to design less complex, stable *per-virtual output queue* scheduling schemes.

The network consists of B switches and each switch has N_b, $1 \leq b \leq B$, inputs and outputs. If the total number of flows in the system is C, we do not have more than N_b^2 physical queues and CN_b^2 logical queues at switch b. We can model the whole network of switches as a system of $\sum_{1 \leq b \leq B} CN_b^2$ logical queues. For the sake of simplicity, we suppose that $N_b = N$, $\forall b$, $1 \leq b \leq B$ and set $K = CN^2B$. Finally, we define $Q_I(b,i)$ as the set of indexes corresponding to the logical queues at the i-th input of the b-switch. Analogously, $Q_O(b,i)$

denotes the set of indexes corresponding to the logical queues directed to the i-th output of the b-switch. We use these definitions to adapt the norm $||Z||_{maxL}$ to a network of switches:

Definition 3. *For $Z \in \mathbb{R}^K, Z = \{z^k, k = CN^2b + CNi + Cj + l, 0 \le b < B, 0 \le i, j < N, 0 \le l < C\}$, the norm $||Z||_{IO}$ is defined as:*

$$||Z||_{IO} = \max_{\substack{b=1,..,B \\ i=1,...,N}} \left\{ \sum_{m \in Q_I(b,i)} |z^m|, \sum_{m \in Q_O(b,i)} |z^m| \right\}.$$

As we assume a deterministic routing policy, the necessary condition for rate stability (5) can be written for a network of switches as follows:

Definition 4. *For a network of IQ/CIOQ switches, a traffic and routing pattern W is admissible if and only if:*

$$||W||_{IO} = ||\Lambda(I - R)^{-1}|| < 1. \tag{6}$$

In the following, we always assume that the condition (6) is satisfied.

3   Maximum Weight Matching Local Scheduling Policies

3.1   Weight Function

All scheduling policies introduced in this paper are matching policies. Any matching policy is defined relative to a specific weight. For the definition of the weights, we will make use of a family of real positive functions $f_k(x) : \mathbb{N} \to \mathbb{R}, 1 \le k \le K$, that satisfy the following property:

$$\lim_{n \to \infty} \frac{f_k(n)}{n} = \frac{1}{w^k} \qquad \text{w.p.1.} \tag{7}$$

We define $\overline{d}^k(n) = \sum_{m \le n} d_m^k$ as the cumulative number of services at queue q^k up to time n. For a given positive constant C, we define the weight of the queue q^k at time n as

$$\phi_n^k = n - f_k(\overline{d}^k(n)) + C. \tag{8}$$

We set

$$\Phi_n = (\phi_n^1, .., \phi_n^K). \tag{9}$$

We see that for a fixed function $f(\cdot)$ that satisfies the relation (7), there exists a constant $C > 0$ such that $\forall n, n \ge 0$, there holds $f(\overline{d}_n^k) \le \frac{\overline{d}_n^k}{w^k} - C$. Further, because the accumulative departure rate at each logical queue q^k cannot be more than the accumulative arrival rate, there holds $\lim_{n \to \infty} \frac{\overline{d}^k(n)}{w^k} \le n$. Combining these two estimates, we see that for any function $f(\cdot)$ that satisfies the condition (7), one can always find a $C > 0$ such that the weights ϕ_n^k are bigger than 1. $\forall n,\, n \geq 0,\, \forall k,\, 1 \leq k \leq K$. In the sequel, we will always assume that the weights ϕ_n^k are bigger than 1.

In [1], an example for $f_k(n)$ is given. The cumulative function of external arrivals for the logical queue q^k is given by $\overline{a}^k(n) = \sum_{m \leq n} a_m^k$. The inverse function $[\overline{a}^k]^{-1}(p)$ maps the packet number p to the arrival slot. Setting $C = 0$ and $f_k(p) = [\overline{a}^k]^{-1}(p)$, the weight

$$\phi_n^k = n - [\overline{a}^k]^{-1}(p) \tag{10}$$

denotes the age of the p-th packet at time n, i.e., it denotes the time the packet has already spent in the network. At its departure time n, the age of the p-th packet is $n - [\overline{a}^k]^{-1}(\overline{d}_n^k)$.

3.2   The Fluid Methodology

We use the fluid methodology and its extension given in [1] and [5]. Based on the definitions introduced in section II , we define the three following vectors: $X(t) = (X_1(t), ..., X_K(t))$ denotes the number of packets in the logical queues at time t, $D = (D_1(t), ..., D_K(t))$ denotes the number of packet departures from the logical queues until time t and $A = (A_1(t), ..., A_K(t))$ denotes the number of packets arrivals at the logical queues until time t. We define $\Pi = \{\pi\}$ as the set of all possible network-wide matchings and denote a specific scheduling algorithm by $\mathcal{F}$. For all $\pi \in \Pi$, we denote by $T_\pi^{\mathcal{F}}(t)$ the cumulative amount of time that the matching π has been used up to time t by the algorithm $\mathcal{F}$. Obviously, $T_\pi^{\mathcal{F}}(0) = 0\, \forall \pi \in \Pi,\, \forall \mathcal{F}$. Using (2), we obtain the fluid equations of the system as follows:

$$X(t) = X(0) + \Lambda t - D(t)(I - R), \tag{11}$$

$$D(t) = \sum_{\pi \in \Pi} \pi T_\pi^{\mathcal{F}}(t), \tag{12}$$

$$\sum_{\pi \in \Pi} T_\pi^{\mathcal{F}}(t) = t. \tag{13}$$

The first equation models the evolution of the logical queues, whereas the second equation counts the total number of departures from the $VOQs$. The third equation reflects the fact that in each timeslot each input is connected to some output.

3.3   Maximum Weight Matching Policies

In this section, we define a class of maximum weight matching policies that guarantee the stability of a network of input-queued switches. Using the fluid methodology, we can define a function $\Phi(t)$ based on the definition of the function Φ_n in (9). We define a set of functions $\mathcal{G}$:

Definition 5. *A real function F is said to belong to the set $\mathcal{G}$ if*
a) F is monotonically non-decreasing, $F(0) = 0$, $F(x) > 0$ if $x > 0$.
b) $\dot{F}(x)$ exists for all $x > 0$.

For a fixed set of functions $F_1, .., F_K \in \mathcal{G}$, we define the following vector function that consists of the derivatives of the functions $F_1, .., F_K$:

$$V_{\dot{F}}(\Phi(t)) = \left(\dot{F}_1(\phi^1(t), ..., \dot{F}_K(\phi^K(t)) \right).$$

We define $\Gamma = [\gamma^{(i,j)}]$ as the diagonal matrix with $\gamma^{(k,k)} = w^k$, and let Γ^{-1} be the inverse of Γ. Further, we write the scalar product for two vectors v_1 and v_2 as $\langle v_1, v_2 \rangle = v_1 v_2^T$. We define a scheduling algorithm MWM^{V_F} as follows: At each time t, the scheduling algorithm MWM^{V_F} chooses the schedule π^{V_F} which is defined by the following equation:

$$\pi^{V_F}(t) = \arg \max_{\pi} \left\{ \langle \pi, V_{\dot{F}}(\phi(t)) \rangle \right\}. \tag{14}$$

We set $H(x) = \Gamma V_F(x)$, $x \in \mathbb{R}^K$ and define the Lyapunov function:

$$G(t) = \langle \mathbb{I}, H(\Phi(t)) \rangle,$$

We formulate the main result of this section:

Theorem 1. *For any set of functions $F_1, .., F_K \in \mathcal{G}$, a network of IQ/CIOQ switches that implements a MWM^{V_F} scheduling policy achieves 100% throughput.*

3.4   Proof of Theorem 1

We note that by (7) $\lim_{t \to \infty} f_k(t) \to t/w^k$, and that $\overline{d}^k(t) \to \infty$ for $t \to \infty$. Thus, $\phi^k(t) \to t - \frac{\overline{d}^k(t)}{w^k} + C$, which implies:

$$\dot{\Phi}(t) = \mathbb{I} - \dot{D}(t)\Gamma^{-1}. \tag{15}$$

We want to show that $\forall t \geq 0$,

$$\Phi(t) \leq B, \tag{16}$$

for a certain constant $B > 0$. We see that if

$$\frac{d}{dt}G(t) \leq 0 \tag{17}$$

$\forall t > 0$, then there holds $G(t) \leq G(0)$, which in turn implies (16) for a certain $B > 0$. Thus, we will show (17) in order to prove (16). Now (17) follows from (12), (13), (14) and (15):

$$\frac{d}{dt}G(t) = \frac{d}{dt}\langle 1, H(\Phi(t))\rangle$$

$$= \sum_{k=1}^{K} w_k \dot{F}_k(\phi^k(t))\dot{\phi}^k(t)$$

$$= \sum_{k=1}^{K} w_k \dot{F}_k(\phi^k(t))(1 - d_k w_k^{-1})$$

$$= \sum_{k=1}^{K} w_k \dot{F}_k(\phi^k(t)) - \sum_{k=1}^{K} d_k \dot{F}_k(\phi^k(t))$$

$$= \langle W, V_{\dot{F}}(\Phi(t))\rangle - \langle D(t), V_{\dot{F}}(\Phi(t))\rangle$$

$$\leq 0.$$

The relationship (16) implies that $0 < t - \frac{\overline{d}^k(t)}{w^k} + C \leq B$. Therefore,

$$\lim_{t\to\infty} \frac{\overline{d}^k(t)}{t} = w^k, \qquad \text{i.e.,} \qquad \lim_{t\to\infty} \frac{D(t)}{t} = W, \qquad \text{w.p.1,} \qquad (18)$$

which corresponds to the rate stability condition (see definition 1) for $X(t)$. $\square$

Corollary 1. *If in (8) the weights are defined as $[\phi_k^n + 1]$ instead of ϕ_k^n, where $[x]$ denotes the biggest integer smaller than or equal to x, then theorem 1 still holds.*

Proof: The proof is nearly identical to the proof of theorem 1.

3.5   Distributed Implementation of the Algorithm

The maximum weight matching algorithms as defined in (14) formulate the scheduling problem as an optimization problem that takes into account all logical queues of the network. This seems to contradict the purpose of the paper to investigate distributed scheduling policies, in which each switch only considers the logical queues at its own *VOQs*. We will now explain that this is not the case.

The maximization in (14) is subject only to the crossbar constraint: In each timeslot, at each switch at most one cell can be sent from one input/to one output. However, a switch configuration at a specific switch does not constrain the choice of the switch configuration at another switch. Thus, we split the weight vector Φ into B weight subvectors $\Phi = (\Phi_1, .., \Phi_B)$, where the subweight vector Φ_b contains the logical queues at the b-th switch. Accordingly, we split the vector V_F in B subvectors $V_F = (V_{F,1}, .., V_{F,B})$. Thus, the maximization in (14) can be written as:

$$\pi^{\mathcal{V}^{\mathcal{F}}}(t) = \arg\max_{\pi}\{\langle \pi, V_{\dot{F}}(\phi(t))\rangle\} = \sum_{b=0}^{B} \arg\max_{\pi_b}\left\{\langle \pi_b, V_{\dot{F},b}(\Phi_b(t))\rangle\right\},$$

where π_b is the schedule chosen at the b-th switch. The maximization problem $\max_{\pi_b} \left\{ \langle \pi_b, V_{\dot{F},b}(\phi_b(t)) \rangle \right\}$ can be solved solely at the b-th switch.

4   Maximal Weight Matching Algorithms

4.1   Definition of a Maximal Weight Matching Algorithm

In this section, we formally define a maximal weight matching algorithm for an arbitrary set of weights. For a given input i and a given output j at a given switch b, we define the set of all logical queues that either belong to the input i or that are directed to the output j. We set $\forall b, i, j, \; 1 \leq b \leq B, \; 1 \leq i, j \leq N$,

$$\mathcal{S}_{b,i,j} := \left\{ m : 1 \leq m \leq K, \; m \in Q_I(b,i) \bigcup Q_O(b,j) \right\}.$$

For a set of positive weights P^k, $1 \leq k \leq K$, where P^k is the weight assigned to the logical queue q^k, we now formally define a maximal weight matching algorithm as follows:

1. Initially, all logical queues q^k, $1 \leq k \leq K$, are considered potential choices for a cell transfer.
2. The logical queue with the largest weight, say q^{k_0}, is chosen for a cell transfer and ties are broken randomly. We assume without loss of generality that $k_0 \in Q_I(b_1, i_1)$ and $k_0 \in Q_O(b_1, j_1)$.
3. All logical queues q^k with $k \in \mathcal{S}_{b_1, i_1 j_1}$ are removed.
4. If all q^k are removed, the algorithm terminates. Else go to step 2.

4.2   A Stable Maximal Weight Matching Algorithm

In this section, we define a specific maximal weight matching algorithm MM^Φ without speedup that guarantees the stability of a network of switches when it is deployed without a speedup. A maximal weight matching algorithm with general weights has been defined in the previous section. Thus, here we only have to define the specific weights of the algorithm MM^Φ. We denote by $p_1 < p_2 < .. < p_K$ the first K prime numbers in increasing order. Using (8), we define the weight w_k of the k-th logical queue as

$$w_k =: exp(v_k) =: exp(p_k[\phi_k + 1]^2),$$

where exp is the exponential function. The definition of the weights ensures that no two weights are of equal value in any timeslot. This is true because for any two different logical queues k and m, i.e, $k \neq m$, the largest power of p_k which divides v_k is odd, whereas the largest power of p_k that divides v_m is even or zero. We now state the main result of this paper:

Theorem 2. *A network of IQ/CIOQ switches that implements the MM^Φ scheduling policy with a speedup $S = 1$ achieves 100% throughput.*

Proof: We follow an idea in [8] to show that the algorithm MM^Φ always calculates the same matching as the following maximum weight matching algorithm: We choose the functions F_k as $F_k(x) = exp(p_k x^2)$. As in corollary 1, we then replace x by $[x + 1]$. We denote the corresponding maximum weight matching policy as defined in (14) by MWM^E. This maximum weight matching algorithm is stable by corollary 1. The stability of MM^Φ then follows from the fact that MM^Φ and MWM^E always calculate the same matching. This is shown using an argument from [8]. In the first iteration, the MM^Φ algorithm chooses the queue with the largest value of all w_k, say w_a. We recall that no two weights w_i and accordingly no two values v_i have the same value. Thus, the MWM^E algorithm, which maximizes the weight of the whole matching, will also choose the queue with the weight $w_a = exp(v_a)$ for packet transfer because

$$\sum_{\substack{k=1 \\ k \neq a}}^{K} exp(v_a) \leq \sum_{k=0}^{v_a - 1} exp(k) = \frac{exp(v_a) - 1}{exp(1) - 1} < exp(v_a).$$

Due to the crossbar structure of the switch, neither the MM^Φ nor the MWM^E algorithm chooses any logical queue for packet transfer that competes for a switch input and output with the chosen logical queue. Thus, these logical queues can be discarded for the rest of the proof. Applying the same arguments to the subset of the remaining queues, one sees that both algorithms choose the queue with the biggest weight w_b among the remaining queues. The successive application of this argument shows that both scheduling algorithms are indeed identical.

5   Conclusions

This paper examines scheduling algorithms with low complexity for networks of CIOQ switches. We propose an algorithm that satisfies the most common performance requirements on a scheduling algorithm: It guarantees the stability of a network of switches, it can be implemented in distributed manner that does not require the exchange of information between the switches, it is of feasible computational complexity and it does not require any speedup.

References

1. Ajmone, M.,Giaccone, P., Leonardi, E., Mellia, M., Neri, F., *Local scheduling policies in networks of packet switches with input queues*, Proc. of Infocom 2003, San Francisco, April 2003.
2. Ajmone, M.,Leonardi, E., Mellia, M., Neri, F., *On the throughput achievable by isolated and interconnected input-queued switches under multicalss traffic*, Proc. of Infocom 2002, New York City, June 2002.
3. Andrews, M., Zhang, L., *Achieving stability in networks of input queued switches* Proc. of Infocom 2001, Anchorage, Alaska, April 2001.

4. Benson, K., *Throughput of crossbar switches using maximal weight matching algorithms,* Proc. of IEEE ICC 2002, New York City.
5. Dai, J.G., Prabhakar, B., *The throughput of data switches with and without speedup,* Proc. of IEEE Infocom 2000, Tel Aviv.
6. Keslassy, I., McKeown, N., *Achieving 100% throughput in an input queued switch,* IEEE Transactions on Communications, vol. 47, no. 8, Aug. 1999, 1260 - 1272.
7. Shah, D., Kopikare, M., *Delay bounds for approximate maximum weight matching algorithms for input queued switches,* Proc. of IEEE Infocom 2002, New York City, June 2002.
8. Shah, D,; *Stable Algorithms for input queued switches,* Proc. 39th Annual Allerton Conference on Communication, Control and Computing, Oct. 2001.

The External Network Problem with Edge- or Arc-Connectivity Requirements*

Jan van den Heuvel and Matthew Johnson

Centre for Discrete and Applicable Mathematics, Department of Mathematics,
London School of Economics, Houghton Street, London WC2A 2AE, UK
{jan, matthew}@maths.lse.ac.uk.

Abstract. The connectivity of a communications network can often be enhanced if the nodes are able, at some expense, to form links using an external network. In this paper, we consider the problem of how to obtain a prescribed level of connectivity with a minimum number of nodes connecting to the external network.

Let $D = (V, A)$ be a digraph. A subset X of vertices in V may be chosen, the so-called external vertices. An internal path is a normal directed path in D; an external path is a pair of internal paths $p_1 = v_1 \cdots v_s$, $p_2 = w_1 \cdots w_t$ in D such that v_s and w_1 are external vertices (the idea is that v_1 can contact w_t along this path using an external link from v_t to w_1). Then (D, X) is externally-k-arc-strong if for each pair of vertices u and v in V, there are k arc-disjoint paths (which may be internal or external) from u to v.

We present polynomial algorithms that, given a digraph D and positive integer k, will find a set of external vertices X of minimum size subject to the requirement that (D, X) must be externally-k-arc-strong.

1 Introduction

Communications networks have an obvious modelling in terms of graphs or digraphs, and often it is possible to express a network's reliability, or endurance, using some graph-theoretic property of the (di)graph that represents it. In this note we consider the following situation : suppose that the nodes of a network N can access an external network. For example, N could be a fairly loose network whose nodes can also communicate using a second, more reliable, network. Another example is a wireless network where additional links can be formed using satellite connections. Suppose further that by using the external network the reliability of N can be improved, but that its use should be minimized as there is an associated cost (this could simply be a financial cost, or it could be a loss of security, or a restriction of node mobility). The problem we have is this :

- if we require N to have a prescribed level of reliability, which nodes should use the external network ?

This is the most general statement of the *External Network Problem*.

---

* Research supported by EPSRC MathFIT grant no. GR/R83514/01.

A. López-Ortiz and A. Hamel (Eds.): CAAN 2004, LNCS 3405, pp. 114–126, 2005.

In this paper we are mainly concerned with the problem when the reliability of the network N is given by the arc-strong connectivity of a digraph $D = (V, A)$. The analogous problem for undirected graphs is significantly easier, and can be found in the next short section.

We need some definitions before we can formally state the problem. An internal path in D is a sequence of vertices and arcs $v_1 e_1 v_2 \cdots e_{t-1} v_t$ where, for $1 \le i \le t-1$, e_i is the arc $v_i v_{i+1}$. The *set of external vertices* is a subset of V and is denoted X_V. An *external path* of D is a pair of internal paths $p_1 = v_1 \cdots v_s$, $p_2 = w_1 \cdots w_t$ where v_s and w_1 are external vertices. Thus, in this paper, when we use the term *path* without specifying whether it is internal or external, it can be either. The digraph D is *externally-k-arc-strong*, and X_V is *k-good*, if between each pair of vertices in V, there are k arc-disjoint paths. A digraph is *internally-k-arc-strong* if each pair of vertices is joined by k-arc-disjoint internal paths (this is just the usual definition of k-arc-strong).

External Network Problem

Input: A digraph $D = (V, A)$ and a positive integer k.
Output: A k-good set of external vertices $X_V \subseteq V$ of minimum size.

In the final section we shall present polynomial algorithms that will find a k-good set X_V of minimum size. In the next section we consider the special case of undirected graphs. In Sections 3 and 4 we introduce some preliminary results needed before we can present the algorithms.

First we discuss an alternative setting for this problem. Suppose D is a digraph which is not k-arc-strong. The (symmetric) *Source Location Problem* requires us to find a smallest possible set S (a subset of V that we call a set of sources) such that for each $v \in V \setminus S$ there are k arc-disjoint internal paths from S to v and k arc-disjoint paths from v to S. We show that a k-good set of external vertices is also a set of sources and vice versa. Let X_V be a k-good set of external vertices for D, and choose a vertex $v \in X_V$. For each $u \in V \setminus X_V$, there are k arc-disjoint paths from u to v. If any of the paths are external, then consider only the first internal path that joins u to an external vertex: this provides k arc-disjoint internal paths from u to X_V. To obtain k arc-disjoint internal paths from X_V to u, take the k arc-disjoint paths from v to u and, for those that are external, consider only the second internal path that joins an external vertex to u. Thus X_V is a set of sources. Now suppose that S is a set of sources. Identify the vertices of S to form a new digraph D'. Clearly D' is internally-k-arc-strong. Thus each pair of vertices are joined by k arc-disjoint internal paths, and, if we let the vertices of S be external vertices, these correspond to arc-disjoint paths in D. So S is a k-good set of external vertices. Ito *et al.* [5] described a polynomial algorithm for the Source Location Problem in the case where k is fixed; the algorithms we present (which would also provide solutions to the Source Location Problem) are polynomial even if k is not fixed.

There are many other ways in which we could interpret the general statement of the problem given in the opening paragraph. For example, if the principal cause of unreliability in a network is node failure, then this can be modelled using vertex-connectivity. We cannot expect that, in general, the Source Location Problem (which has been extensively studied) will provide solutions to the External Network Problem : the Source Location Problem with vertex-connectivity requirements (see [4, 6]) is not equivalent to the natural formulation of the External Network Problem with vertex-connectivity requirements. This is discussed further in a companion paper [2].

It is also possible to generalize the External Network Problem with arc-connectivity requirements for digraphs: let $c : V(D) \longrightarrow \mathbb{R}$ be a cost function and let $d : V(D) \times V(D) \longrightarrow \mathbb{R}$ be a demand function. Now a set of external vertices is *good* if for each pair of vertices $u, v \in V$ there are $d(u, v)$ arc-disjoint (internal or external) paths from u to v, and the problem is to find a minimum cost set of good external vertices. We have shown that this generalization is NP-hard if either the cost or demand function is non-uniform; there is not space in this paper to present our proof of this.

2   Undirected Graphs

In this section $G = (V, E)$ is an undirected graph and k is a positive integer. The following definitions are the obvious analogues of those for digraphs : a set of external vertices X_V is a subset of V; an internal path of G is a sequence of incident vertices and edges and an external path is pair of internal paths where the final vertex of the first and the initial vertex of the second are external vertices; G is externally-k-edge-connected, and X_V is k-good, if each pair of vertices is joined by k edge-disjoint paths.

We shall describe how to find a minimum size k-good set of external vertices for G. This could be achieved by replacing each edge of G by a pair of oppositely oriented arcs and using the algorithms for digraphs presented in the final section. We shall see, however, that the problem for undirected graphs is far simpler than that for digraphs.

For $P \subseteq V$, $d(P)$ is the number of edges joining P to $V \setminus P$.

Definition 1. *A non-empty set $P \subseteq V$ is k-critical if $d(P) < k$, and, for each proper subset Q of P, $d(Q) \geq k$.*

It can easily be seen that X_V is a k-good set of external vertices if and only if it intersects each k-critical set. We say that X_V *covers* any k-critical set that it intersects. Thus our aim is to cover all the k-critical sets of G with as few vertices as possible. We need the following lemma, which is well-known and easily proved.

Lemma 2. *Let P and Q be subsets of V. Then $d(P) + d(Q) \geq d(P \setminus Q) + d(Q \setminus P)$*

Proposition 3. *The k-critical sets of G are pairwise disjoint.*

Proof. Let P and Q be distinct k-critical sets of G. By the definition of k-critical, one cannot be a subset of the other. So if P and Q are not disjoint, then $P \setminus Q \neq P$ and $Q \setminus P \neq Q$. Using the definition of critical again,

$$d(P \setminus Q) + d(Q \setminus P) \geq 2\,k, \qquad \text{and}$$
$$d(P) + d(Q) < 2\,k.$$

This contradicts Lemma 2 which proves that the hypothesis that P and Q are not disjoint must be false. □

We immediately have the following result.

Theorem 4. *A minimum size k-good set of external vertices for a graph G contains as many vertices as there are k-critical sets in G.*

It is not necessary, however, to find the k-critical sets to find a minimum size k-good set of external vertices. To check that a set X_V is k-good, identify the vertices of X_V to form a graph G': X_V is k-good if and only if G' is k-edge-connected which can be checked using, for example, a flow algorithm. To find a minimum k-good set, let $X_V = V$ and then repeatedly remove any vertex from X_V as long as the set obtained is k-good. In this way a minimal set is obtained, and every minimal set is a set of minimum size.

3   Critical Sets in Digraphs

In the sequel, $D = (V, A)$ is a digraph and k is a positive integer.

For $P \subseteq V$, $d^-(P)$ is the number of arcs joining $V \setminus P$ to P, and $d^+(P)$ is the number of arcs joining P to $V \setminus P$.

Definition 5. *A non-empty set $P \subseteq V$ is k-critical if $d^-(P) < k$ or $d^+(P) < k$, and, for each proper subset Q of P, $d^-(Q) \geq k$ and $d^+(Q) \geq k$.*

A k-critical set P is k-in-critical if $d^-(P) < k$ and k-out-critical if $d^+(P) < k$ (it is possible for a set to be both k-in-critical and k-out-critical).

Again, it can easily be seen that X_V is a k-good set of external vertices if and only if it intersects each k-critical set.

Definition 6. *Let $\mathcal{P} = P_1, \ldots, P_s$ be a collection of k-critical sets. The relation graph of these sets has vertex set $\mathcal{P}$ and contains an edge joining P_i and P_j, $i \neq j$, if $P_i \cap P_j \neq \varnothing$.*

We will say that k-critical sets are *neighbours* if they intersect (even when not explicitly referring to a relation graph).

We investigate the structure of relation graphs. The results of this section follow easily from the results of Ito *et al.* [5] on the Source Location Problem.

Lemma 7. *If P is a k-in-critical set in D and $Q \neq P$ is a k-out-critical set in D, then $P \cap Q = \varnothing$.*

Proof. By the definition of k-critical neither P nor Q is a subset of the other. If they are not disjoint, then the three sets of vertices $S_1 = P \setminus Q$, $S_2 = P \cap Q$ and $S_3 = Q \setminus P$ are all non-empty. Let $S_4 = V \setminus (P \cup Q)$ (this is possibly the empty set), and let $s_{i,j}$ be the number of arcs from S_i to S_j. As $d^-(S_1) \geq k$ and $d^+(S_3) \geq k$ (since they are proper subsets of k-critical sets),

$$s_{2,1} + s_{3,1} + s_{4,1} \geq k, \tag{1}$$
$$s_{3,1} + s_{3,2} + s_{3,4} \geq k. \tag{2}$$

And as $P = S_1 \cup S_2$ is in-critical and $Q = S_2 \cup S_3$ is out-critical,

$$s_{3,1} + s_{3,2} + s_{4,1} < k, \tag{3}$$
$$s_{2,1} + s_{3,1} + s_{3,4} < k. \tag{4}$$

Adding (1) to (2), and (3) to (4), we obtain a contradiction.   □

Recall that an undirected graph G is *chordal* if it contains no induced cycles of length more than three (that is, if every cycle of length at least four has a *chord*, an edge joining two vertices that are not adjacent in the cycle).

Proposition 8. *For any digraph, the relation graph G of any collection of its k-critical sets is a chordal graph and if a collection of k-critical sets $\mathcal{Q}$ form a clique in G, then $\bigcap\limits_{P \in \mathcal{Q}} P \neq \varnothing$.*

Proof. If G is not chordal, there is a cycle induced by some sets $P_1, \ldots, P_t, t \geq 4$. By Lemma 7, these sets are either all in-critical or all out-critical. Suppose that the latter holds (it will be clear that the former case can be similarly proved). We can assume that $P_i \cap P_j \neq \varnothing$ if $|i - j| = 1 \bmod t$ and $P_i \cap P_j = \varnothing$ otherwise. Then

$$\sum_{|i-j|=1 \bmod t} d^+(P_i \cap P_j) \leq \sum_{i=1}^{t} d^+(P_i) < tk.$$

The first sum is obtained by counting the number of arcs from $P_i \cap P_j$ to $V \setminus (P_i \cap P_j)$. Each such arc joins P_i to $V \setminus P_i$ or P_j to $V \setminus P_j$ (or both) so is also counted at least once when evaluating the second sum. The second inequality follows from the definition of k-critical. Thus, as there are t terms in the first sum, $d^+(P_i \cap P_j) < k$ for some $i, j, |i - j| = 1 \bmod t$, a contradiction as $P_i \cap P_j$ is a proper subset of a k-critical set. This proves that G is chordal.

If $\mathcal{Q}$ is a clique of size 1 or 2, then there is a vertex in every set in $\mathcal{Q}$. Let $\mathcal{Q} = P_1, \ldots, P_t, t \geq 3$, be the smallest clique such that $\bigcap\limits_{i=1}^{t} P_i = \varnothing$. Again we can assume that the k-critical sets are all out-critical. Then

$$\sum_{i=1}^{t} d^+\left(\bigcap_{j \neq i} P_j\right) \leq \sum_{i=1}^{t} d^+(P_i) < tk.$$

Again, every arc counted when evaluating the first sum is counted at least once for the second sum (the second inequality is the same as before). Each of the t sets in the first sum is non-empty (as each intersection is of a collection of sets that forms a clique smaller than $\mathcal{Q}$, so they have a common vertex), and $d^+\left(\bigcap\limits_{j\neq i} P_j\right) < k$ for some i, a contradiction as these sets are proper subsets of k-critical sets. Therefore there must be a vertex contained in every set in $\mathcal{Q}$, and the proposition is proved.   □

Theorem 9. *A minimum size k-good set of external vertices for a digraph D is the same size as a maximum size family of disjoint k-critical sets.*

Proof. A k-good set of external vertices must be at least as big as a family of disjoint k-critical sets since it must intersect each set in the family. We prove the theorem by finding a k-good set of external vertices $X_V = \{x_1, \ldots, x_t\}$ and a family of disjoint k-critical sets $P_1, \ldots, P_t$ such that x_i covers P_i, $1 \le i \le t$.

We shall use a well-known property of chordal graphs (see, for example, [1]).

- For any chordal graph G, there exists $v \in V(G)$ such that v and its neighbours form a clique.

Therefore, using this property and Proposition 8,

- for any collection $\mathcal{P}$ of k-critical sets, there exists a k-critical set $P \in \mathcal{P}$ such that the intersection of P and all its neighbours in $\mathcal{P}$ is non-empty; we call P an *end-set*.

Let G_1 be the relation graph of all the k-critical sets of D. Let P_1 be an end-set of G_1, and let x_1 be a vertex that covers P_1 and all its neighbours. Let G_2 be the relation graph of all the k-critical sets not covered by x_1 and note that it contains no k-critical set that intersects P_1. Now suppose that we have found $x_1, \ldots, x_s$ and $P_1, \ldots, P_s$, $s < t$, and that the relation graph G_{s+1} of k-critical sets not yet covered contains no critical set that intersects P_i, $1 \le i \le s$. Let P_{s+1} be an end-set of G_{s+1} and let x_{s+1} be a vertex that covers P_{s+1} and all its neighbours. Note that G_{s+2}, the relation graph of uncovered k-critical sets, contains no set that intersects P_i, $1 \le i \le s+1$. If G_{s+2} is the null graph, then $s + 1 = t$ and we are done.   □

4   External Subsets

The algorithms will use a generalization of sets of external vertices: a *set of external subsets* is a disjoint collection of non-empty sets $\mathcal{X}_V = \{X_1, \ldots, X_t\}$ such that $X_i \subseteq V$ for $1 \le i \le t$.

Definition 10. *A set of external subsets is k-good if $\bigcup\limits_{i=1}^{t} X_i$ is a k-good set of external vertices.*

The remaining definitions in this section assume that $\mathcal{X}_V$ is a k-good set of external subsets. An external subset $X \in \mathcal{X}_V$ is *redundant* if $\mathcal{X}_V \setminus \{X\}$ is also k-good. If $\mathcal{X}_V$ contains no redundant set, it is *minimally k-good*.

Definition 11. *For $u \in V$ and $X \in \mathcal{X}_V$, if $(\mathcal{X}_V \setminus \{X\}) \cup \{\{u\}\}$ is also a k-good set of external subsets, then u is an* alternative *to X. The* unrestricted set of alternatives *to X contains all such u and is denoted $A(X)$. The* restricted set of alternatives *to X is $A(X) \cap X$ and is denoted $B(X)$.*

In the algorithms, a common operation is to alter $\mathcal{X}_V$ by replacing one of the subsets X by its (restricted or unrestricted) set of alternatives X'. Notice that if $X' \neq \varnothing$, then $(\mathcal{X}_V \setminus \{X\}) \cup \{X'\}$ is also k-good.

Definition 12. *For $X \in \mathcal{X}_V$, a k-critical set is an* essential set *of X if it is covered by X but not by any other external subset.*

A vertex is an alternative to X if and only if it covers its essential sets. Thus if X is not redundant, $A(X)$ is equal to the intersection of the essential sets of X; if X is redundant, it has no essential sets and $A(X) = V$.

Definition 13. *If a set $X \subseteq V$ is a subset of an k-critical set P, then P is a* confining set *of X. If X is equal to the intersection of its confining sets, then it is* confined.

If a set $X \in \mathcal{X}_V$ is not redundant, then $A(X)$ is confined by the essential sets of X. If X is confined, then $B(X)$ is confined by the confining sets of X and the essential sets of X.

Definition 14. *$\mathcal{X}_V$ is* stable *if for each $X \in \mathcal{X}_V$, $X = A(X)$. It is* consistent *if for each $X \in \mathcal{X}_V$, $X = B(X)$.*

Notice that if $\mathcal{X}_V$ is stable, then it is also consistent and each $X \in \mathcal{X}_V$ is confined.

Proposition 15. *If X_V is a minimum size k-good set of external vertices and $\mathcal{X}_V$ is a stable set of external subsets, then $|X_V| \geq |\mathcal{X}_V|$.*

In the next section, we will see that from a stable set, it is possible to find a minimum size k-good set that contains one vertex from each external subset. This will prove that $|X_V| \leq |\mathcal{X}_V|$. Thus $|X_V| \geq |\mathcal{X}_V|$

Proof. Let $\mathcal{X}_V = \{X_1, \ldots, X_t\}$. We will prove that $|X_V| \geq |\mathcal{X}_V|$ by finding a disjoint collection of k-critical sets $P_1, \ldots, P_t$.

As each external subset in $\mathcal{X}_V$ is its own set of alternatives, it is equal to the intersection of its essential sets, and each of these essential sets intersects only one source-set.

We use the structure of relation graphs of k-critical sets of D. Consider the relation graph G_1 of the essential sets of all of the sets $X_1, \ldots, X_t$. Let P_1 be an end-set of this graph, and, we might as well assume, that P_1 is an essential set for X_1. Thus $X_1 = A(X_1) \subseteq P_1$, and the essential sets of X_1 are P_1 and none, some or all of its neighbours. Recall from Proposition 8 that the intersection of P_1 with all its neighbours is non-empty. Each vertex in this intersection is certainly an alternative to X_1 — it covers all the essential sets — and thus a member of X_1. Therefore, as the essential sets each intersect only one of the source-sets, P_1 and *all* of its neighbours in G_1 are essential sets of X_1.

Now consider the relation graph G_2 of the essential sets except those of X_1. Let P_2 be an end-set and suppose it is an essential set for X_2. Note that P_2 does not intersect P_1 as no essential set that intersects P_1 is included in G_2. By the same argument as before, P_2 and all of its neighbours in G_2 are essential sets of X_2.

Then we look for an end-set in the relation graph of all essential sets except those of X_1 and X_2. From this and further repetitions we find $P_3, \ldots, P_t$.   □

When we present the algorithms in the next section, it will be assumed that it is possible to check that a set of external vertices X_V is k-good in polynomial time. This can be done by contracting the vertices of X_V to obtain a digraph D' and checking that D' is k-arc-strong using, for example, a flow algorithm. Furthermore, we can check that $\mathcal{X}_V$ is a minimally k-good by checking that $\mathcal{X}_V \setminus \{X\}$ is not k-good for each $X \in \mathcal{X}_V$. To find the set of alternatives to some $X \in \mathcal{X}_V$, first check whether or not $\mathcal{X}_V \setminus \{X\}$ is k-good: if it is, then $A(X) = V$; otherwise $A(X)$ contains each vertex $u \in V$ such that $(\mathcal{X}_V \setminus \{X\}) \cup \{\{u\}\}$ is k-good.

5   Algorithms

We present two polynomial algorithms: STABLESUBSETS finds a stable set of external subsets for a digraph D, and MINIMUMSET takes a stable set and finds a set of external vertices containing a single vertex from each external subset. By Proposition 15, this set of external vertices will have minimum size.

Using these algorithms, the time needed to find a minimum size k-good set of external vertices for a digraph $D = (V, A)$ is $O(n^4 m \log(n^2/m))$ where $n = |v|$ and $m = |A|$. The bottleneck is the loop labelled L1: it can be shown that this loop takes time $O(n^3 S(n))$, where $S(n)$ is the complexity of an algorithm that decides whether a set of vertices is k-good. As we remarked at the end of the previous section this can be done using a flow algorithm; in particular an algorithm of Hao and Orlin [3] means we can take $O(n m \log(n^2/m))$ for $S(n)$.

In the remainder of this section we prove the efficacy of the two algorithms.

To begin, STABLESUBSETS considers the vertex set of V as a set of external subsets $\mathcal{X}_V$. Redundant sets are discarded until $\mathcal{X}_V$ is a minimal k-good set of external subsets.

The main part of the algorithm contains three nested **while** loops labelled L1, L2 and L3. We say that the algorithm *enters* a loop if the *loop condition* is satisfied. For example, the loop condition for L2 is that $\mathcal{X}_V$ contains a redundant set. Inside the loops $\mathcal{X}_V$ is altered by replacing external subsets by their sets of alternatives or by discarding external subsets. We shall show that after each alteration $\mathcal{X}_V$ is still k-good. Thus if the algorithm does not enter L1, then for each $X \in \mathcal{X}_V$, $X = A(X)$. Hence $\mathcal{X}_V$ is a k-good stable set of external subsets and we have the required output.

Algorithm STABLESUBSETS

Input: A digraph $D = (V, A)$ where $V = \{v_1, \ldots, v_n\}$.
Output: A k-good stable set of external subsets $\mathcal{X}_V$ for D.

let $\mathcal{X}_V = \{\{v_1\}, \ldots, \{v_n\}\}$;
while there exists a redundant set $R \in \mathcal{X}_V$ **do**
  let $\mathcal{X}_V = \mathcal{X}_V \setminus \{R\}$;
end /* while */
while there exists $Y \in \mathcal{X}_V$, $Y \neq A(Y)$ **do**   /* L1 */
  let $\mathcal{X}_V = (\mathcal{X}_V \setminus \{Y\}) \cup \{A(Y)\}$;
  while there exists a redundant set $R \in \mathcal{X}_V$ **then**   /* L2 */
    let $\mathcal{X}_V = \mathcal{X}_V \setminus \{R\}$;
    while there exists $Z \in \mathcal{X}_V$, $Z \neq B(Z)$ **do**   /* L3 */
      let $\mathcal{X}_V = (\mathcal{X}_V \setminus \{Z\}) \cup \{B(Z)\}$.
    end /* while */
  end /* while */
end /* while */

Output $\mathcal{X}_V$.

Algorithm MINIMUMSET

Input: A digraph $D = (V, A)$ with a k-good stable set of external subsets $\mathcal{X}_V$.
Output: X_V, a minimum size k-good set of external vertices for D.

while there exists $R' \in \mathcal{X}_V$, $|R'| \geq 2$ **do**   /* L4 */
  choose $u \in R'$;
  let $\mathcal{X}_V = (\mathcal{X}_V \setminus \{R'\}) \cup \{\{u\}\}$;
  while there exists $Z \in \mathcal{X}_V$, $Z \neq B(Z)$ **do**   /* L5 */
    let $\mathcal{X}_V = (\mathcal{X}_V \setminus \{Z\}) \cup \{B(Z)\}$.
  end /* while */
end /* while */
let $X_V = \bigcup_{X \in \mathcal{X}_V} X$.

Output X_V.

We shall prove the following stronger claims. (Recall that a set of external subsets $\mathcal{X}_V$ is consistent if $X = B(X)$ for each $X \in \mathcal{X}_V$.)

Claim. (a) Each time the algorithm considers the loop condition for L1, $\mathcal{X}_V$ is a consistent k-good set of external subsets, and each $X \in \mathcal{X}_V$ is confined or a singleton, but not redundant.
(b) Each time the algorithm considers the loop condition for L2, $\mathcal{X}_V$ is a consistent k-good set of external subsets, and each $X \in \mathcal{X}_V$ is confined or a singleton.

We prove the claim by induction. The first time the algorithm considers the loop condition for L1, $\mathcal{X}_V$ is a set of singletons and minimal. Thus each external subset is non-redundant and its own restricted set of alternatives, and hence $\mathcal{X}_V$ is consistent.

Suppose that $X = B(X)$ and X is non-redundant for each $X \in \mathcal{X}_V$ and that L1 is entered. A set Y is replaced by $Y' = A(Y)$ (and $Y \subsetneqq Y'$ since $Y = B(Y) \subseteq A(Y) \neq Y$). Clearly $B(Y') = Y'$, Y' is confined, and $\mathcal{X}_V$ is k-good. For each $X \in \mathcal{X}_V$, $X \neq Y'$, X remains a singleton or confined, and also the set of alternatives to X will, if anything, have grown when Y was replaced by $A(Y)$. Thus for all $X \in \mathcal{X}_V$, the claim that X is confined or a singleton and $X = B(X)$ still holds when the loop condition for L2 is considered. So if L2 is not entered, none of the external subsets is redundant; the algorithm returns to consider the condition for L1 and Claim 5 (a) holds.

Suppose that L2 is entered and a redundant set R is removed from $\mathcal{X}_V$. If L3 is not entered, then Claim 5 (b) holds.

So suppose that L3 is entered. We call the process encoded by L3 *reduction* : to *reduce* a k-good set of external subsets is to arbitrarily choose a external subset X and replace it by $B(X)$, and to repeat this until a k-good consistent set is obtained (note that after each alteration, the restricted sets of alternatives of each external subset must be recalculated). A k-good set of external subsets is *reducible* if this process is possible, that is, if each time we replace a external subset X by $B(X)$ we obtain a k-good set of external subsets (clearly the process must eventually terminate as the external subsets are continually getting smaller).

Proposition 16. *Let $\mathcal{X}_V$ be a consistent k-good set of external subsets such that each $X \in \mathcal{X}_V$ is confined or a singleton. Let $Z \in \mathcal{X}_V$ and let $Z' = \varnothing$ or $Z' = \{z\}$ for some $z \in Z$. If $(\mathcal{X}_V \setminus \{Z\}) \cup \{Z'\}$ is a k-good set of external subsets, then it is reducible. Moreover, each X in the set of external subsets obtained after the reduction is confined or a singleton.*

The proof is left to the end of the section.

Apply the proposition to the set $\mathcal{X}_V$ obtained just before R is discarded with $Z = R$ and $Z' = \varnothing$. Thus after the algorithm has finished looping through L3, Claim 5 (b) holds. Then another redundant set may be discarded and the algorithm may begin to loop through L3 again. When no more redundant sets can be found, the algorithm has finished its run through L1 and Claim 5 (a) holds. We have shown that STABLESUBSETS will output a k-good stable set of external subsets.

MINIMUMSET contains two nested **while** loops, L4 and L5. If the algorithm does not enter L4, then each external subset contains one vertex and from this we obtain a minimum size k-good set of external vertices.

As before, we must show that as $\mathcal{X}_V$ is altered, it is always a k-good set of external subsets.

Claim. Each time the algorithm considers the loop condition for L4, $\mathcal{X}_V$ is a k-good consistent set of external subsets, and each $X \in \mathcal{X}_V$ is confined or a singleton.

We use induction to prove the claim. The first time the algorithm considers the loop condition for L4, $\mathcal{X}_V$ is stable and the claim holds (see the remark after Definition 14). Assume the claim holds as L4 is entered. After replacing a external subset R' by one of its alternatives, $\mathcal{X}_V$ is still k-good (by Definition 11). If the algorithm does not enter L5, then the claim is true. All that remains to be proved is that once the algorithm finishes going through L5, the resulting $\mathcal{X}_V$ satisfies the conditions in the claim. Notice that L5 is identical to L3: so applying Proposition 16 with $Z = R'$ and $Z' = \{u\}$ will guarantee that this is the case. We have shown that MINIMUMSET will output a minimum size k-good set of external vertices.

We require one further result on chordal graphs before we prove Proposition 16.

Lemma 17. *Let u and v be non-adjacent vertices in a chordal graph G, and suppose that W_1, the set of vertices adjacent to both u and v, is non-empty. Let W_2 contain each vertex (other than u and v) that is adjacent to every vertex in W_1. Then u and v are in different components in $G - (W_1 \cup W_2)$.*

Proof. We show that if there is a u-v path in $H = G - (W_1 \cup W_2)$, then there is an induced cycle of length greater than 3 in G.

Let $up_1 \cdots p_r v$ be the shortest u-v path in H ($r \geq 2$ as $p_1 \notin W_1$). As $p_1 \notin W_2$, there exists $w \in W_1$ such that p_1 and w are not adjacent in G. If possible, choose the smallest $i \geq 2$ such that p_i is adjacent to w in G: then $\{u, p_1, \ldots, p_i, w\}$ induces a cycle in G. If no p_i is adjacent to w, then $\{u, p_1, \ldots, p_r, v, w\}$ induces a cycle. $\qquad\square$

Proof of Proposition 16. Let $\mathcal{X}_V$, Z and Z' be as in the proposition. During the reduction of $(\mathcal{X}_V \setminus \{Z\}) \cup \{Z'\}$ we repeatedly select an external subset X to replace with its restricted set of alternatives $B(X)$. A singleton will never be selected as it is equal to its restricted set of alternatives. If X is confined, then $B(X)$ is also confined, and if $B(X) \neq \varnothing$, then replacing X by $B(X)$ will give a new set of external subsets. Hence the proposition holds if $B(X) \neq \varnothing$ for all external subsets X encountered during reduction. Thus we will assume that we find a external subset with an empty restricted set of alternatives and show that this leads to a contradiction.

First some terminology: if a set X' is obtained from X by any number of replacements, then we say that X is an *ancestor* of X' and X' is a *descendant* of X (a set is its own ancestor and descendant).

Every external subset obtained during reduction is a subset of a external subset of $\mathcal{X}_V$. Thus if X and X^* are unrelated confined external subsets (neither is the ancestor of the other), then they are disjoint and there is a confining set of X that does not intersect all the confining sets of X^* (since if the relation graph of the confining sets of X and X^* were a clique, there would, by Proposition 8, be a vertex contained in all of them, hence in $X \cap X^*$). So there is a confining set of X that does not intersect X^*.

Now we assume that X is the first external subset obtained during reduction with $B(X) = \varnothing$, and find a contradiction. We can assume that $|X| \geq 2$ and X is confined. Thus $B(X)$ is equal to the intersection of the confining sets of X and the essential sets of X. As $B(X) = \varnothing$, by Proposition 8, two of these sets, say V_1 and V_2, are disjoint (and V_1 and V_2 must be essential sets of X, as X is a subset of each of its confining sets). Let G be the relation graph of all the in-critical sets of D. Apply Lemma 17 with $u = V_1$ and $v = V_2$ (W_1 is non-empty as it includes all the confining sets of X): we obtain a graph $H = G - (W_1 \cup W_2)$ such that V_1 and V_2 are in separate components of H, say H_1 and H_2.

We shall find a path in H from H_1 to H_2, a contradiction. We need the following result.

Assertion 18. *Suppose that an external subset Y is replaced by $B(Y)$ during reduction and that P is an essential but not confining set of Y. Then either*

- *$P \cap Z \neq \varnothing$, or*
- *there is a external subset T that was replaced by $B(T)$ earlier during reduction (i.e., before Y is replaced by $B(Y)$), and $P \cap T \neq \varnothing$ but $P \cap B(T) = \varnothing$.*

Proof. Let Y' be the ancestor of Y in $\mathcal{X}_V$. As P is not a confining set of Y, it is not a confining set of Y'. Thus P was covered by another external subset in $\mathcal{X}_V$ (else $Y' \neq B(Y')$, contradicting that $\mathcal{X}_V$ is consistent). If this was Z, then $P \cap Z \neq \varnothing$. Otherwise suppose a external subset W covered P. When Y is replaced by $B(Y)$, no descendant of W covers P. Thus at some point before Y was replaced, a descendant of W that does cover P was replaced by its restricted set of alternatives that does not cover P. Let this descendant be T. $\square$

We use the assertion to find a sequence $X_1 P_1 X_2 \cdots X_r P_r$ where,

- $X_1 = X$ and $P_1 = V_1$;
- for $1 \leq j \leq r$, X_j is a confined external subset, P_j is an essential but not confining set of X_j;
- for $1 \leq j \leq r - 1$, P_j is covered by X_{j+1} but not by $B(X_{j+1})$;
- $P_r \cap Z \neq \varnothing$;
- sets that are not consecutive in the sequence are disjoint.

The first two terms of the sequence are given. When the first $2j$ terms, $X_1 P_1 X_2 \cdots X_j P_j$, are known, apply Assertion 18 with $Y = X_j$ and $P = P_j$. If $P_j \cap Z \neq \varnothing$, then the sequence is found. Otherwise let $X_{j+1} = T$. Since X was the first external subset encountered with an empty restricted set of alternatives, $B(X_{j+1}) \neq \varnothing$ and is confined. As $B(X_{j+1})$ does not cover P_j, X_{j+1} must have an essential set that does not cover P_j: let this set be P_{j+1}. Note that P_{j+1} is not a superset of X_j and thus not a confining set of it. We must show that X_{j+1} and P_{j+1} are each disjoint from the sets they are not consecutive to in the sequence. By the choice of P_{j+1} and X_{j+1}, $P_j \cap P_{j+1} = \varnothing$; and X_{j+1} is unrelated to X_j so they are disjoint. Let Q_j and Q_{j+1} be disjoint confining sets of X_j and X_{j+1}, respectively. Then $P_{j+1} \cap X_j = \varnothing$, else $\{Q_j, P_j, Q_{j+1}, P_j\}$ induces a 4-cycle in G. If $j > 1$, we must show that, for $1 \leq j' \leq j - 1$, X_{j+1} and P_{j+1} do not intersect $X_{j'}$ or $P_{j'}$. Apply Lemma 17 with $u = P_j$ and $v = P_{j-1}$ (W_1 is not empty as it includes all the confining sets of X_j): we obtain a graph $J = G - (W_1 \cup W_2)$ such that P_j and P_{j-1} are in separate components of J, say J_1 and J_2. As Q_{j+1} covers P_j and does not intersect X_j, it must be in J_1. For $1 \leq j' \leq j - 1$, let $Q_{j'}$ be a confining set of $X_{j'}$ that does not intersect X_j.

Note that $P_{j-1}Q_{j-1}P_{j-2}\cdots P_1 Q_1$ is a path in G and must be in J_2 since none of these in-critical sets intersect X_j. Thus, for $1 \le j' \le j-1$, P_{j+1} and Q_{j+1} (and so X_{j+1}) are disjoint from $P_{j'}$ and $Q_{j'}$.

Every time in the construction above we find a set X_j that is replaced by its restricted set of alternatives $B(X_j)$ during reduction earlier than X_{j-1} was replaced by $B(X_{j-1})$. Therefore after a finite number of steps the sequence must end with a suitable P_r.

Once the sequence is found, let Q'_j, $2 \le j \le r$, be a confining set of X_j that does not intersect $X = X_1$. Then $P_1 Q'_2 P_2 \cdots Q'_r P_r$ must be a path in H_1. Thus we have found an in-critical set P_r in H_1 that intersects Z.

Use the same argument to find an in-critical set W in H_2 that intersects Z (find a path from V_2). If $|Z| \ge 2$, then there is a confining set U of Z that does not intersect every confining set of X. Thus $P_{r-1}UW$ is a path in H from H_1 to H_2. If $|Z| = 1$, then $P_{r-1} \cap W \supseteq Z$. Hence $P_{r-1}W$ is an edge in H. This final contradiction completes the proof of Proposition 16.   □

References

1. R. Diestel, *Graph Theory*. Springer-Verlag, New York, 1997.
2. J. van den Heuvel and M. Johnson, *The External Network Problem with vertex-connectivity requirements*. In preparation.
3. J. Hao and J. B. Orlin, *A faster algorithm for finding the minimum cut in a graph*. J. of Algorithms, **17** (1994), 424–446.
4. H. Ito, M. Ito, Y. Itatsu, H. Uehara, and M. Yokohama, *Location problems based on node-connectivity and edge-connectivity between nodes and node-subsets*. Lecture Notes in Computer Science, **1969** (2000), 338–349.
5. H. Ito, K. Makino, K. Arata, S. Honami, Y. Itatsu, and S. Fujishige, *Source location problem with flow requirements in directed networks*. Optimization Methods and Software, **18** (2003), 427–435.
6. H. Nagamochi, T. Ishii, and H. Ito, *Minimum cost source location problem with vertex-connectivity requirements in digraphs*. Info Process Letters, **80** (2001), 287–294.

Bipartite Graphs as Models
of Complex Networks

Jean-Loup Guillaume and Matthieu Latapy

LIAFA – CNRS – Université Paris 7,
2 place Jussieu, 75005 Paris, France
{guillaume, latapy}@liafa.jussieu.fr

Abstract. We propose here the first complex network model which achieves the following challenges: it produces graphs which have the three main wanted properties (clustering, degree distribution, average distance), it is based on some real-world observations, and it is sufficiently simple to make it possible to prove its main properties. This model consists in sampling a random bipartite graph with prescribed degree distribution. Indeed, we show that any complex network can be viewed as a bipartite graph with some specific characteristics, and that its main properties can be viewed as consequences of this underlying structure.

1   Introduction

It has been shown recently that most real-world complex networks have some essential properties in common. These properties are unfortunately not captured by the model generally used before this discovery, although they play a central role in many contexts like the robustness of the Internet, the spread of viruses or rumors over the Internet, the Web or others social networks, as well as the performance of protocols and algorithms. This is why, in the last few years, a strong effort has been put in the realistic modeling of complex networks and much progress has been accomplished is this field.

In this paper, we propose the random bipartite graph model as a general model for complex networks. It produces graphs with all the wanted properties. It relies on real-world observations and gives realistic graphs. Finally, it is simple enough to make it possible to prove its main properties.

Most real-world complex networks have a number of edges m which scales linearly with the number of vertices n: $m \sim k \cdot n$ where k is the average degree which does not depend on the size of the graph. Moreover, three other properties received recently much attention: the average distance between vertices, the clustering coefficient (probability of existence of an edge between two vertices when they are both neighbors of a same vertex) and the degree distribution (for each k, the probability p_k that a randomly chosen vertex has degree k). It has been shown that for almost all real-world complex networks the average distance is small (it scales as $\log(n)$), the clustering is high (independent of the size of

A. López-Ortiz and A. Hamel (Eds.): CAAN 2004, LNCS 3405, pp. 127–139, 2005.

Table 1. The main statistics for the complex networks we use in this paper. For each network, we give its number of vertices n, its number of links m, the value of the exponent α of its power law degree distribution, its clustering coefficient C, and its average distance d. Moreover, we give the values of these statistics for typical graphs with the same number of vertices and links obtained using the purely random model (C_{rd} and d_{rd}), the random model with prescribed degree distribution (C_{dd} and d_{dd}), the Albert and Barabási model (C_{ab} and d_{ab}) and the one of Watts and Strogatz (C_{ws} and d_{ws}). Recall that purely random graph and Watts and Strogatz ones have a Poisson degree distribution, which makes the α exponent irrelevant in these cases. Moreover, in the cases pointed by a star (*), the real clustering coefficient is too large to be obtained with the Watts and Strogatz model. Therefore we used in these cases the parameters inducing the maximal clustering, *i.e.* no rewiring, which yields very large average distances

	Internet	Web	Actors	Co-auth	Co-occur	Protein
n	228263	325729	392340	16401	9297	2113
m	320149	1090108	15038083	29552	392066	2203
α	2.3	2.3	2.2	2.4	1.8	2.4
C	0.06	0.466	0.785	0.638	0.822	0.153
C_{rd}	0.00001	0.00002	0.0002	0.0002	0.009	0.001
C_{dd}	0.001	0.017	0.0057	0.001	0.26	0.007
C_{ab}	0.0002	0.0005	0.0015	0.003	0.028	0
C_{ws}	0.06	0.461	0.74 (*)	0.523 (*)	0.74 (*)	0.06 (*)
d	9.2	7	3.6	7.18	2.13	6.74
d_{rd}	11.96	5.47	2.97	7.57	2.06	10.4
d_{dd}	5.6	4.48	2.95	5.77	2.36	5.73
d_{ab}	7.2	5.1	2.93	5.5	2.38	8.15
d_{ws}	15.6	11.23	2559 (*)	2269 (*)	55.6 (*)	509 (*)

the graph) and the degree distribution follows a power law ($p_k \sim k^{-\alpha}$, with α generally between 2 and 3).

The basic model for complex networks is the random graph model, but many other attempts have been made recently to provide more realistic models. The most used ones are the Erdös and Rényi model which yields graphs with a low diameter but no clustering and a Poisson degree distribution; the Molloy and Reed model which has both degree distribution and diameter; the Watts and Strogatz and Albert which has both diameter and clustering; and finally the Albert and Barabási model which misses the clustering. We will compare our model to these four models in this paper. For definitions and surveys of these models and their main properties, we refer to [2, 5, 13].

In this paper, we propose a solution to the random sampling of graphs which have all the three wanted properties. To achieve this, we focus on another property of *all* real-world complex networks, namely their underlying bipartite structure (Section 2). We then propose two models: the random sampling of bipartite graphs with prescribed degree distributions, and the growing bipartite model with preferential attachment (Section 3). Indeed, as shown in Sections 4 and 5, respectively formally and experimentally, these models induce the three wanted properties. This means that these can be viewed as consequences of the underlying bipartite structure of all complex networks, which is our main contribution.

Throughout our presentation, we will use a representative set of complex networks which have received much attention and span quite well the variety of context in which complex networks appear. The set consists of a protein interaction graph [9], a map of the Internet at the router level [8], the map of a large Web site [3] (considered as undirected), the actors' co-starring relation [17], the co-occurrence relation of words [6] in the sentences of the Bible [16], and a co-authoring relation between scientists [12]. We will refer to these graphs as *Proteins*, *Internet*, *Web*, *Actors*, *Cooccurrence*, and *Coauthoring* respectively. They are precisely defined and studied in the cited references.

The main properties of these real-world complex networks are given in Figure 1. Values obtained with the main models we have cited are also provided. Notice that, as announced, all these real-world complex networks have a very low average distance, a power law distribution of degrees, and a high clustering. Graphs obtained using the models are significantly different concerning at least one of these three points.

2   Complex Networks as Bipartite Graphs

A bipartite graph is a triple $G = (\top, \bot, E)$ where $\top$ and $\bot$ are two disjoint sets of vertices, respectively the top and bottom vertices, and $E \subseteq \top \times \bot$ is the set of edges of the graph. The difference with classical (unipartite) graphs lies in the fact that edges exist only between top and bottom vertices.

Given a bipartite graph $G = (\top, \bot, E)$, one can easily obtain its unipartite version defined as $G' = (\bot, E')$ where $\{u, v\}$ is in E' if u and v are both connected to a same (top) vertex in G. See Figure 1. In this unipartite version of the graph, each top vertex induces a clique (complete subgraph) between the bottom vertices to which it is linked.

Some complex networks display a natural bipartite structure. For instance, one can view *Actors* (two actors are linked if they are part of a same cast) as a bipartite graph where $\top$ is the set of movies, $\bot$ is the set of actors, and each

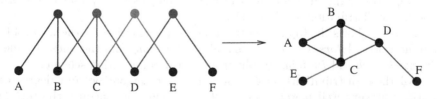

Fig. 1. A bipartite graph and its unipartite version. Notice that the link $\{B, C\}$ is obtained twice since B and C have two neighbors in common in the bipartite graph

actor is linked to the movies he/she played in. *Coauthoring* and *Cooccurrence* can also be described this way.

However, most complex networks do not display such a structure. For example there is no immediate and natural way to see *Internet*, *Web* or *Proteins* as bipartite graphs. And yet, we can check experimentally that these networks contain large cliques: between 20 and 30 vertices in *Internet*, 100 and more vertices in *Web*. The existence of large cliques in these graphs makes it interesting to describe them as bipartite graphs as follows: the top vertices are the cliques contained in the graph we consider, and the bottom vertices are the vertices of the graph itself. A clique and a vertex are linked if the vertex belongs to the clique.

We still have to specify which cliques we consider. Given a graph $G' = (V, E')$, we want to obtain a set of cliques $C = \{C_i\}$ such that G' is the unipartite version of the bipartite graph $G = (C, V, E)$ where $E = \{\{C_i, v\} | v \in C_i\}$. This problem can also be viewed as follows: we look for a set of cliques such that the edges in E are exactly the edges in the cliques. This problem is known as the *clique covering problem* [15].

A trivial solution is given by $C = E'$: each clique (of size 2) covers exactly one edge of the graph. However, our aim is to obtain a set of cliques C such that the bipartite graph G will have properties similar to the ones observed for natural bipartite graphs: large cliques should be discovered and the number of cliques should be kept linear in the size of the graph. Computing maximal cliques is NP-complete [1,7] as well as minimizing the number of cliques [11]. However, some heuristics make it possible to compute such cliques if the graph is not too large. In our case, we use the following remarks. Let us denote the sets of neighbors of a vertex and a edge by $N(u) = \{v \in V | \{u, v\} \in E\}$ and $N(u, v) = N(u) \cap N(v)$ respectively. First notice that a largest clique containing $\{u, v\}$ in G' is also a largest clique containing $\{u, v\}$ in the sub-network of G' induced by $N(u, v) \cup \{u, v\}$.

In real-world complex networks, we observed that the subgraphs induced by $N(u, v)$ for all edges $\{u, v\}$ are in general very dense and very small. This makes it possible to compute the following clique covering of these complex networks: for each edge $\{u, v\}$ in G' we take the largest clique containing it (if there are more than one, we choose one at random). We obtain this way a number of cliques bounded by the number of edges in the graph. Moreover, the large cliques of the graph will be discovered. We can check experimentally that the decomposition of a bipartite graph in cliques returns a set of cliques very similar to the original one (same distribution of cliques size).

Figure 2 shows the top and bottom degree distribution for the natural bipartite networks *Actors*, *Cooccurrence* and *Coauthoring*, and the ones obtained for *Internet*, *Web* and *Proteins* graphs using our decomposition scheme.

All these distributions have a property in common: bottom degree distributions fit very well power laws in all cases. On the contrary, the top degree distributions are of two kinds: while *Cooccurrence*, *Coauthoring*, *Internet* and *Proteins* ones exhibit a Poisson behavior, *Actors* and *Web* ones are more heavy

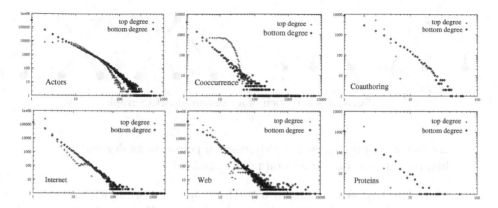

Fig. 2. Top and bottom degree distributions for the natural bipartite versions of *Actors*, *Cooccurrence*, and *Coauthoring*, and for the bipartite version of *Internet*, *Web*, and *Proteins* obtained with the decomposition scheme

tailed. These results lead to several remarks. First, the power law bottom degree distribution seems universal, just like the power law distribution in the unipartite versions of these graphs. Second, the top degree distributions can be qualitatively different. We do not enter in more details in this extended abstract, but this point is important in the use of the bipartite structure for modeling complex networks since it can impact on some characteristics of the generated graphs. Further remarks will be pointed in Section 6.

The fact that *all* complex networks have a nontrivial underlying bipartite structure, which can be computed using our decomposition scheme, leads us to the following question: is it possible to see the main properties of real-world complex networks as consequences of their underlying bipartite structure? We answer this question in the next sections.

3   The Models

Our aim is now to use the new general property of real-world complex networks discovered in the previous section, namely their underlying bipartite structure, as a way to propose a model which captures the main wanted properties. As discussed in the first sections of this paper, there are basically two ways to achieve this goal. First, we may try to sample random bipartite graphs with prescribed (top and bottom) degree distributions. Second, we may try to propose a construction process similar to the ones observed in practice, to obtain a *growing* model.

One can sample uniformly a random bipartite graph with prescribed $\top$ and $\bot$ degree distributions as follows (see Figure 3) [4, 14]:

1. generate both top and bottom vertices and assign to each vertex a degree drawn from the given distributions,

Fig. 3. Construction of a random bipartite graph

2. create for each vertex as many connection points as its degree,
3. link top and bottom connection points randomly,

This process generates random bipartite graphs uniformly within the set of bipartite graphs with the given degree distribution. However few constraints must be considered while generating the graph [14].

The random bipartite model assumes that two distributions, for both top and bottom degrees, are explicitly given. We can also use the preferential attachment principle to define them implicitly and introduce a growing model. Indeed, as already noticed, the bottom degree distributions follow a power law. This leads us to the following model: at each step, a new top vertex is added and its degree d is sampled from a prescribed (top) distribution (which qualitatively varies between various graphs). Then, for each of the d edges of the new vertex, either a new bottom vertex is added (with probability $1 - \lambda$) or we pick one among the preexisting ones using preferential attachment (with probability λ). The parameter λ is the *overlap ratio*, defined as the average ratio of preexisting bottom vertices to which a new top vertex is connected. It can be measured on average on real-world bipartite graphs, which gives values between 0.75 and 0.95.

At each step the bipartite graph has the required degree distributions. Notice that this construction process is very similar to the one observed in some real-world cases. For instance, *Actors* is built exactly this way: when a new movie is produced (which corresponds to the addition of a top vertex), it is linked to actors according to their popularity, and to some new actors, playing in a movie for the first time.

We finally have two models to produce bipartite networks similar to the ones obtained from real-world complex networks, in terms of top and bottom degree distributions. The others properties of interest (average distance, degree distribution and clustering) are also captured with this model as shown formally and with experimental results in the next sections.

4   Analysis of the Models

Our aim in this section is to give sketches of the proofs for the main properties of the unipartite version of a random bipartite graph with prescribed degree distributions (see annex for complete proofs). Since these properties are induced by a *typical* graph (this is what random sampling gives us), this is a way to answer the following question: what properties are induced by the underlying bipartite structure?

Degree Distribution

Given a bottom vertex u, we denote by $d(u)$ the degree of u in the bipartite graph, and by $d_U(u)$ its degree in the unipartite graph. We want to study the distribution of $d_U(u)$.

Lemma 1. *Let us consider a bottom vertex $u \in \bot$. The number of bottom vertices which have a neighbor (in $\top$) in common with u, i.e. $d_U(u)$, is:*

$$\frac{d(u)}{|\top|} \cdot \sum_{t \neq u} d(t) + \mathcal{O}\left(\frac{d(u)^2}{|\top|^2} \cdot \sum_{t \neq u} d(t)^2\right)$$

If the bottom degree distribution is a power law with exponent β, then :

$$P[d_U(u) = k] \sim P[d(u) = \frac{n}{\sum_{t \neq u} d(t)} \cdot k] \sim \frac{1}{(\sum_{t \neq u} d(t)) \cdot k)^\beta} \sim k^{-\beta}$$

Therefore, as long as the bottom degree distribution follows a power law, the degree distribution in the unipartite version of the graph also follows a power law with the same exponent, which is indeed the case in practice as one can check in Figures 2 and 4.

Average Distance

To study the average distance in the unipartite version of a graph obtained with the model, we use a result from L. Lu [10] about the diameter (*i.e.* the largest distance between any two vertices) of some specific random graphs. We can deduce that a bipartite graph whose bottom degree distribution follows a power law with an exponent greater than 2, have an average distance (in the unipartite version of G) which is almost surely $\mathcal{O}(\log(|\bot|))$.

Clustering Coefficient

Hereafter we give a lower bound for the clustering coefficient of a graph G' which is the unipartite version of a bipartite graph $G = (\top, \bot, E)$ obtained using the random bipartite model. This bound is valid under reasonable assumptions on the top and bottom degree distributions and is independent of the size of the graph.

First notice that the probability for two top vertices to have more than just one bottom vertex in common in their neighborhood tends to zero when the size of the graph grows. We therefore consider any vertex b in the unipartite version of the graph and we suppose that its neighborhood is composed of a set of disjoint cliques. We can prove the following two lemmas:

Lemma 2. *Let $T_{>2}$ denote the set of top neighbors of b in G with degree strictly greater than 2, and $U_{>2}$ denote the set of bottom neighbors of $T_{>2}$. Let p be the fraction of neighbors of b in G' which belong to $U_{>2}$, and α be the clustering coefficient of b restricted to $U_{>2}$.*

Then the clustering coefficient of b in G' scales as $p^2 \cdot \alpha$.

Therefore, as long as p is constant (which is the case for the distributions met in practice), one can neglect the top vertices of degree 2 when computing the clustering coefficient of a given vertex.

Lemma 3. *If b is connected only to top vertices of degree at least 3 in G, then* $cc(b) \geq \frac{1}{2 \cdot d(b) - 1}$

The clustering coefficient of G' can now be easily approximated: $cc(G') \sim \frac{1}{N} \sum_{b \in \perp} \frac{1}{2d(b)-1}$. As long as there is a linear number $c \cdot N$ of vertices b of degree 2 (this holds in particular for power law networks), the sum scales linearly with N, therefore the clustering coefficient is larger than a non-zero constant independent of the size of the graph.

Finally, we gave here formal proofs of the fact that the random bipartite graph model, with reasonable prescribed degree distributions, produces graphs having the three main wanted properties. Notice that other properties of interest, like the way the obtained graphs are connected, can be studied formally. However, we cannot develop these results in this extended abstract. Notice also that in [14] the authors study similar questions with techniques from statistical physics. We obtain similar and new results which give additional information on the objects in concern. However, their approach is very interesting and the two papers may be considered as two complementary studies.

5   Experimental Results

The formal results of the previous section give a precise intuition on how the random bipartite graph model with prescribed degree distributions behaves. We can also check its properties experimentally by generating graphs using this model and the same parameters as the ones measured on real-world complex networks. This is what we do in this section with our six examples, for the purely random bipartite model, as well as for the one with preferential attachment.

Figure 2 gives the values obtained for the average distance and the clustering coefficient. Figure 4 shows a comparison between the degree distributions of the original graphs, and the ones obtained with the two bipartite models.

As expected from the previous section, the graphs we obtain with the random bipartite model have a power law distribution of degrees, a small average distance and a high clustering coefficient. Moreover, by definition, they have the same distribution of cliques size as the original network. Therefore the model is qualitatively accurate for the modeling of general real-world complex networks. Notice however that the clustering coefficient for *Internet* is badly approximated by the model, the reason of this behavior rely on the existence of a very dense (but not complete) core in *Internet*.

These experimental results should also be compared to the ones obtained with the currently most used models, presented in Section 1, see Figure 1. This

Table 2. The original clustering C and average distance d of our six examples, together with the ones obtained with the random bipartite model with the same degree distributions (C_{rb} and d_{rb}) and with the growing bipartite model (C_{gb} and d_{gb})

	Internet	Web	Actors	Co-auth	Co-occur	Protein
C	0.060	0.466	0.785	0.638	0.822	0.153
C_{rb}	0.456	0.663	0.767	0.542	0.831	0.187
C_{gb}	0.346	0.708	0.793	0.632	0.768	0.244
d	9.2	7	3.6	7.18	2.13	6.74
d_{rb}	4.33	3.2	3.06	5.07	2.06	5.8
d_{gb}	4.59	3.53	2.83	3.98	2.6	5.45

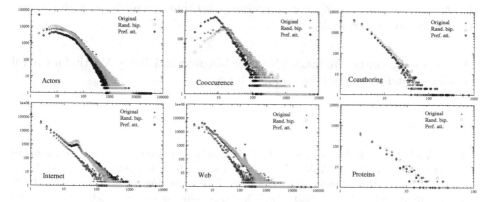

Fig. 4. The original degree distribution of our six examples, together with the ones obtained with the random bipartite model and with the growing bipartite model

comparison gives evidence for the fact that the models we propose may be considered as relevant choices for the realistic modeling of complex networks.

All these remarks hold both for the growing bipartite model and for the random one. This is worth to notice, since it may be very important in some contexts that the model produces *growing* graphs with realistic properties, and in other contexts that the obtained graphs are representative of a whole class of graphs.

Finally, simulation results can be considered as an experimental confirmation of the formal results of previous section, as well as strong indicators that the underlying bipartite structure captures the three main properties observed in practice.

6   Conclusion and Discussion

In this paper, we have proposed a complex network model which achieves the following challenges:

- it has the three main wanted properties (logarithmic average distance, high clustering and power law degree distribution),
- it is based on a *realistic* construction process representative of what happens for some real complex networks, and
- its definition is simple enough to make it possible to give some intuition and some proofs of its properties.

The model is based on the discovery that all real-world complex networks have an underlying bipartite structure which can be seen as responsible for their main properties. Some networks naturally have this structure. For others, we show that they can be decomposed into cliques which make such a structure emerge. This shows that the main properties of complex networks can be viewed as consequences of this bipartite structure, and that the model captures a very general behavior of complex systems.

Acknowledgments. We thank Clémence Magnien and James Martin for careful reading of preliminary versions and useful comments.

References

1. J. Abello, P.M. Pardalos, and M.G.C. Resende. On maximum clique problems in very large graphs. In AMS-DIMACS Series on Discrete Mathematics and Theoretical Computer Science, editors, *External Memory Algorithms*, volume 50, 1999.
2. R. Albert and A.-L. Barabási. Statistical mechanics of complex networks. *Reviews of Modern Physics 74, 47*, 2002.
3. R. Albert, H. Jeong, and A.-L. Barabási. Diameter of the world wide web. *Nature*, 401:130–131, 1999.
4. Source code for the random bipartite graph generator. http://www.liafa.jussieu.fr/~guillaume/programs/.
5. S.N. Dorogovtsev and J.F.F. Mendes. Evolution of networks. *Adv. Phys. 51, 1079-1187*, 2002.
6. R. Ferrer and R.V. Solé. The small-world of human language. In *Proceedings of the Royal Society of London*, volume B268, pages 2261–2265, 2001.
7. M. Garey and D. Johnson. *Computers and Intractability; A Guide to the Theory of NP-Completeness.* W. H. Freeman and Company, 1979.
8. R. Govindan and H. Tangmunarunkit. Heuristics for internet map discovery. In *IEEE INFOCOM 2000*, pages 1371–1380, Tel Aviv, Israel, March 2000. IEEE.
9. H. Jeong, B. Tombor, R. Albert, Z. Oltvai, and A.-L. Barabási. The large-scale organization of metabolic networks. *Nature, 407, 651*, 2000.
10. L. Lu. The diameter of random massive graphs. In ACM-SIAM, editor, *12th Ann. Symp. on Discrete Algorithms (SODA)*, pages 912–921, 2001.
11. S.D. Monson, N.J. Pullman, and R. Rees. A survey of clique and biclique coverings and factorizations of (0,1)-matrices. *Bull. Inst. Combin. Appl.*, 14:17–86, 1995.
12. M.E.J. Newman. Scientific collaboration networks: I. Network construction and fundamental results. *Phys. Rev. E*, 64, 2001.
13. M.E.J. Newman. The structure and function of complex networks. *SIAM Review*, 45(2):167–256, 2003.

14. M.E.J. Newman, D.J. Watts, and S.H. Strogatz. Random graph models of social networks. *Proc. Natl. Acad. Sci. USA*, 99 (Suppl. 1):2566–2572, 2002.
15. J. Orlin. Contentment in graph theory: Covering graphs with cliques. *Indigationes Mathematicae*, 80:406–424, 1977.
16. Bible Today New International Version. http://www.tniv.info/bible/.
17. D.J. Watts and S.H. Strogatz. Collective dynamics of small-world networks. *Nature*, 393:440–442, 1998.

Annex: Theorems and Lemmas Proofs

Degree Distribution

Proof. **Lemma 4.1:** The exact value of $d_U(u)$ is given by:

$$d_U(u) = \sum_{t \neq u} \left(1 - \frac{\binom{|T|-d(u)}{d(t)}}{\binom{|T|}{d(t)}} \right)$$

since the probability that a given bottom vertex t has a top neighbor in common with u depends only on the degree of both vertices and the number of top vertices. To simplify this formula, we can approximate the ratio $\binom{|T|-d(u)}{d(t)} / \binom{|T|}{d(t)}$ as follows:

$$\frac{\binom{|T|-d(u)}{d(t)}}{\binom{|T|}{d(t)}} = \frac{(|T| - d(u))!(|T| - d(t))!}{|T|!(|T| - d(u) - d(t))!}$$

$$\sim \frac{(|T| - d(u))^{d(t)}}{|T|^{d(t)}}$$

$$\sim 1 - \frac{d(t)d(u)}{|T|} + \mathcal{O}\left(\left(\frac{d(t)d(u)}{|T|} \right)^2 \right)$$

Therefore:

$$d_U(u) \sim \sum_{t \neq u} \left(\frac{d(t)d(u)}{|T|} + \mathcal{O}\left(\left(\frac{d(t)d(u)}{|T|} \right)^2 \right) \right)$$

$$\sim \frac{d(u)}{|T|} \sum_{t \neq u} d(t) + \mathcal{O}\left(\frac{d(u)^2}{|T|^2} \sum_{t \neq u} d(t)^2 \right)$$

Average Distance

Theorem 1 (L. Lu [10]). *Let $G = (V, E)$ be a graph whose vertices are weighted with weights $w_1, \cdots, w_n$, such that each edge $\{i, j\}$ appears with probability $w_i \cdot w_j \cdot p$. If the degrees of the vertices in V follow a power law with an exponent β strictly greater than 2, then the diameter of the graph G is almost surely $\Theta(\log(n))$.*

This theorem, together with the one presented above on the degree distribution of the unipartite version of the graph, leads to the following result:

Theorem 2. *Let $G = (\top, \bot, E)$ be a bipartite graph such that the bottom degree distribution follows a power law with an exponent greater than 2, then the average distance of the unipartite version of G is almost surely $\mathcal{O}(\log(|\bot|))$.*

Proof. Given two bottom vertices u and v in $\bot$, the probability that they are connected in the unipartite version is equal to the probability that they are both linked to a same top vertex in G. This probability is exactly proportional to $d_\bot(u) \cdot d_\bot(v)$. Therefore we can apply Theorem 1 considering that the weight of each vertex is its degree and so the connection probability is ensured, and as long as bottom degree distribution follows a power law with an exponent β strictly greater than 2. The diameter of the unipartite version of the graph is almost surely $\mathcal{O}(\log(|\bot|))$, which is an upper bound for the average distance, which also scales logarithmically.

Clustering Coefficient

Proof. **Lemma 4.2:** The fact that the clustering coefficient of b restricted to $U_{>2}$ is α implies that $|\triangle_{U_{>2}}(b)| = \alpha \cdot \binom{p \cdot d}{2}$. If we consider the whole neighborhood of b, the number of triangles does not change while the number of connected triples increases:

$$cc(b) = \frac{\alpha \cdot \binom{p \cdot d}{2}}{\binom{d}{2}} = \alpha \cdot \frac{p((p \cdot d - 1)}{d - 1} \sim p^2 \cdot \alpha$$

Proof. **Lemma 4.3:** Suppose b is connected to two top vertices, t_1 and t_2, of degree at least 3 (we deal with the general case below). Then the clustering coefficient of b is:

$$cc(b) = \frac{\binom{d(t_1)-1}{2} + \binom{d(t_2)-1}{2}}{\binom{d(t_1)+d(t_2)-2}{2}}$$

Suppose now that b is connected to t_2 and t_1' such that $d(t_1') = d(t_1) + 1$, then the clustering coefficient of b is:

$$cc'(b) = \frac{\binom{d(t_1)+1-1}{2} + \binom{d(t_2)-1}{2}}{\binom{d(t_1)+d(t_2)-1}{2}}$$

and:

$$cc'(b) - cc(b) = \frac{2 \cdot (d(t_2) - 1)}{(d(t_1) + d(t_2) - 2) \cdot (d(t_1) + d(t_2) - 3)} > 0$$

which means that the clustering coefficient grows with the degree of t_1 and t_2. A lower bound for the clustering coefficient of b can therefore be obtained when both t_1 and t_2 have the smallest possible degree, 3.

This can be extended to the case where b has more than two top neighbors to obtain the following lower bound:

$$cc(b) = \frac{\sum_{t_i} \binom{d(t_i)-1}{2}}{\binom{\sum_{t_i}(d(t_i)-1)}{2}} \geq \frac{\sum_{t_i} \binom{3-1}{2}}{\binom{\sum_{t_i}(3-1)}{2}} \geq \frac{1}{2 \cdot d(b) - 1}$$

Traceroute-Like Exploration of Unknown Networks: A Statistical Analysis

Luca Dall'Asta[1], Ignacio Alvarez-Hamelin[1], Alain Barrat[1],
Alexei Vázquez[2], and Alessandro Vespignani[1]

[1] Laboratoire de Physique Théorique (UMR du CNRS 8627),
Bâtiment 210, Université Paris Sud, 91405 Orsay, France
[2] Department of Physics, University of Notre Dame,
Notre Dame, IN 46556, USA

Abstract. Mapping the Internet generally consists in sampling the network from a limited set of sources by using `traceroute`-like probes. This methodology has been argued to introduce uncontrolled sampling biases that might produce statistical properties of the sampled graph which sharply differ from the original ones. Here we explore these biases and provide a statistical analysis of their origin. We derive a mean-field analytical approximation for the probability of edge and vertex detection that allows us to relate the global topological properties of the underlying network with the statistical accuracy of the sampled graph. In particular we show that shortest path routed sampling allows a clear characterization of underlying graphs with scale-free topology. We complement the analytical discussion with a throughout numerical investigation of simulated mapping strategies in different network models.

1 Introduction

A significant research and technical challenge in the study of large information networks is related to the lack of highly accurate maps providing information on their basic topology. This is mainly due to the dynamical nature of their structure and to the lack of any centralized control resulting in a self-organized growth and evolution of these systems. A prototypical example of this situation is faced in the case of the physical Internet. The topology of the Internet can be investigated at different granularity levels such as the router and Autonomous System (AS) level, with the final aim of obtaining an abstract representation where the set of routers or ASs and their physical connections (peering relations) are the vertices and edges of a graph. In the absence of accurate maps, researchers rely on a general strategy that consists in acquiring local views of the network from several points and merge these views in order to get a presumably accurate global map. Local views are obtained by evaluating a certain number of paths to different destinations by using specific tools such as `traceroute` or by the analysis of BGP tables. At first approximation these processes amount to the collection of shortest paths from a source node to a set of target nodes, obtaining a partial

A. López-Ortiz and A. Hamel (Eds.): CAAN 2004, LNCS 3405, pp. 140–153, 2005.
© Springer-Verlag Berlin Heidelberg 2005

spanning tree of the network. The merging of several of these views provides the map of the Internet from which statistical properties are evaluated.

By using this strategy a number of research groups have generated maps of the Internet [1,2,3,4,5], that have been used for the statistical characterization of the network properties. Defining $\mathcal{G} = (V, E)$ as the sampled graph of the Internet with $N = |V|$ vertices and $|E|$ edges, it is quite intuitive that the Internet is a *sparse* graph in which the number of edges is much smaller than in a complete graph; i.e. $|E| \ll N(N-1)/2$. Equally important is the fact that the average distance between vertices, measured as the shortest path, is very small. This is the so called *small-world* property, that is essential for the efficient functioning of the network. Most surprising is the evidence for the heavy-tailed behavior of the degree distribution that has been collected in several studies at the router and AS level [6,7,8,9,10,11] and has generated a large activity in the field of network modeling and characterization [12,13,14,15,16]. Namely, the probability that any vertex in the graph has degree k is well approximated by $P(k) \sim k^{-\gamma}$ with $2 \leq \gamma \leq 2.5$ [6].

While `traceroute`-driven strategies are very flexible and can be feasible for extensive use, the obtained maps are undoubtedly incomplete. Along with technical problems such as the instability of paths between routers and interface resolutions [17], typical mapping projects are run from relatively small sets of sources whose combined views are missing a considerable number of edges and vertices [11, 18]. Such spurious effects, called *sampling biases*, might seriously compromise the statistical accuracy of the sampled graph and have been explored in numerical experiments of various synthetic graphs [19, 20, 21]. Very interestingly, it has been shown that apparent degree distributions with heavy-tails may be observed even from homogeneous topologies such as in the classic Erdös-Rényi graph model [19,20]. These studies thus point out that the evidence obtained from the analysis of the Internet sampled graphs might be insufficient to draw conclusions on the topology of the actual Internet network.

In this work we tackle this problem by a *mean-field* statistical analysis and extensive numerical experiments of shortest path routed sampling in different networks models. We find an approximate expression for the detection probability of edges and vertices that exploits the dependence upon the number of sources, targets and the topological properties of the networks. This expression allows the understanding of the qualitative behavior of the efficiency of the exploration methods by changing the number of probes imposed to the graph. Moreover, the analytical study provides a general understanding of which kind of topologies yields the most accurate sampling. In particular, we show that the map accuracy depends on the underlying network betweenness distribution; the broader the distribution the higher the statistical accuracy of the sampled graph. We substantiate our analytical finding with a throughout analysis of maps obtained varying the number of source-target pairs on networks models with different topological properties. A detailed discussion of the behavior of the degree distribution and other statistical quantities is provided for the various sampled graphs and compared with the insight obtained by analytical means.

2   Modeling the traceroute Discovery of Networks

In a typical traceroute study, a set of active sources deployed in the network run traceroute probes to a set of destination nodes. Each probe collects information on all the nodes and edges traversed along the path connecting the source to the destination, allowing the discovery of the network [17]. By merging the information collected on each path it is then possible to reconstruct a partial map of the network. More in detail, the edges and nodes discovered by each probe will depend on the metric $\mathcal{M}$ used to decide the path between a pair of nodes. While in the Internet many factors, including commercial agreement and administrative routing policies, contribute to determine the actual path, it is clear that to a first approximation the route obtained by traceroute-like probes is the shortest path between the two nodes. This assumption, however, is not sufficient for a proper definition of a traceroute model in that equivalent shortest paths between two nodes may exist. In the presence of a degeneracy of shortest paths we must therefore specify the metric $\mathcal{M}$ by providing a resolution algorithm for the selection of shortest paths.

For the sake of simplicity we can define three selection mechanisms defining different $\mathcal{M}$-paths that may account for some of the features encountered in Internet discovery: (i) *Unique Shortest Path* (USP) probe. In this case the shortest path route selected between a node i and the destination target T is always the same independently of the source S (the path being initially chosen at random among all the equivalent ones). (ii) *Random Shortest Path* (RSP) probe. The shortest path between any source-destination pair is chosen randomly among the set of equivalent shortest paths. This might mimic different peering agreements that make independent the paths among couples of nodes. (iii) *All Shortest Paths* (ASP) probe. The metric discovers all the equivalent shortest paths between source-destination pairs. This might happen in the case of probing repeated in time (long time exploration), so that back-up paths and equivalent paths are discovered in different runs. Actual traceroute probes contain a mixture of the three mechanisms defined above. We do not attempt, however, to account for all the subtleties that real studies encounters, i.e. IP routing, BGP policies, interface resolutions and many others. Each traceroute probe provides a test of the possible biases and we will see that the different metrics have only little influence on the general picture emerging from our results. On the other hand, it is intuitive to recognize that the USP metric represents the worst case scenario since, among the three different methods, it yields the minimum number of discoveries. For this reason, if not otherwise specified, we will report the USP data to illustrate the general features of our synthetic exploration.

More formally, the setup for our simulated traceroute mapping is the following. Let $G = (V, E)$ be a sparse undirected graph with vertices (nodes) $V = \{1, 2 \cdots, N\}$ and edges (links) $E = \{i, j\}$. Then let us define the sets of vertices $\mathcal{S} = \{i_1, i_2, \cdots, i_{N_S}\}$ and $\mathcal{T} = \{j_1, j_2, \cdots, j_{N_T}\}$ specifying the random placement of N_S sources and N_T targets. For each ensemble of source-target pairs $\Omega = \{\mathcal{S}, \mathcal{T}\}$, we compute with the metric $\mathcal{M}$ the path connecting each source-target pair. The sampled graph $\mathcal{G} = (V^*, E^*)$ is defined as the set of vertices V^* (with $N^* = |V^*|$) and edges E^* induced by considering the union of all the $\mathcal{M}$-paths connecting the source-target pairs. The sampled graph is thus analogous to the maps obtained from real `traceroute` sampling of the Internet.

In general, `traceroute`-driven studies run from a relatively small number of sources to a much larger set of destinations. For this reason, in our study the parameters of interest are the number of sources N_S, and the *density* of targets $\rho_T = N_T/N$. Indeed, while 100 targets may represent a fair probing of a network composed by 500 nodes, this number would be clearly inadequate in a network of 10^6 nodes. The density of targets ρ_T thus allows us to compare mapping processes on networks with different sizes by defining an intrinsic percentage of targeted vertices. In many cases, as we will see in the next sections, an appropriate quantity representing the level of sampling of the networks is

$$\epsilon = \frac{N_S N_T}{N} = \rho_T N_S, \tag{1}$$

that measures the density of probes imposed to the system. In real situations it represents the density of `traceroute` probes in the network and therefore a measure of the load provided to the network by the measuring infrastructure.

In the following, our aim is to evaluate to which extent the statistical properties of the sampled graph $\mathcal{G}$ depend on the parameters of our experimental setup and are representative of the properties of the underlying graph G.

3   Mean-Field Theory of the Discovery Bias

We begin our study by presenting a mean-field statistical analysis of the simulated `traceroute` mapping. Our aim is to provide a statistical estimate for the probability of edge and node detection as a function of N_S, N_T and the topology of the underlying graph. For each set $\Omega = \{\mathcal{S}, \mathcal{T}\}$ we can define the quantities

$$\sum_{t=1}^{N_T} \delta_{i,j_t} = \begin{cases} 1 & \text{if vertex } i \text{ is a target;} \\ 0 & \text{otherwise,} \end{cases} \quad \sum_{s=1}^{N_S} \delta_{i,i_s} = \begin{cases} 1 & \text{if } i \text{ is a source;} \\ 0 & \text{otherwise,} \end{cases} \tag{2}$$

where $\delta_{i,j}$ is the Kronecker symbol. These quantities tell us if any given node i belongs to the set of sources or targets, and obey the sum rules $\sum_i \sum_{t=1}^{N_T} \delta_{i,j_t} = N_T$ and $\sum_i \sum_{s=1}^{N_S} \delta_{i,i_s} = N_S$. Analogously, we define the quantity $\sigma_{i,j}^{(l,m)}$ that assumes the value 1 if the edge (i,j) belongs to the $\mathcal{M}$-path between nodes l and m, and 0 otherwise. By using these definitions, the indicator function that a given edge (i,j) will be discovered and belongs to the sampled graph is

$$\pi_{i,j} = 1 - \prod_{l \neq m} \left(1 - \sum_{s=1}^{N_S} \delta_{l,i_s} \sum_{t=1}^{N_T} \delta_{m,j_t} \sigma_{i,j}^{(l,m)} \right). \tag{3}$$

In the case of a given set $\Omega = \{\mathcal{S}, \mathcal{T}\}$, the discovery indicator function is simply $\pi_{i,j} = 1$ if the edge (i,j) belongs to at least one of the $\mathcal{M}$-paths connecting the source-target pairs, and 0 otherwise. While the above exact expression does not lead us too far in the understanding of the discovery probabilities, it is interesting to look at the process on a statistical ground by studying the average over all possible realizations of the set $\Omega = \{\mathcal{S}, \mathcal{T}\}$. By definition we have that

$$\left\langle \sum_{t=1}^{N_T} \delta_{i,j_t} \right\rangle = \rho_T \quad \text{and} \quad \left\langle \sum_{s=1}^{N_S} \delta_{i,i_s} \right\rangle = \rho_S, \tag{4}$$

where $\langle \cdots \rangle$ identifies the average over all possible deployment of sources and targets Ω. These equalities simply state that each node i has, on average, a probability to be a source or a target that is proportional to their respective densities. In the following, we will make use of an uncorrelation assumption that allows an explicit approximation for the discovery probability. The assumption consists in neglecting correlations originated by the position of sources and targets on the discovery probability by different paths. While this assumption does not provide an exact treatment for the problem it generally conveys a qualitative understanding of the statistical properties of the system. In this approximation, the average discovery probability of an edge is

$$\langle \pi_{i,j} \rangle = 1 - \left\langle \prod_{l \neq m} \left(1 - \sum_{s=1}^{N_S} \delta_{l,i_s} \sum_{t=1}^{N_T} \delta_{m,j_t} \sigma_{i,j}^{(l,m)} \right) \right\rangle \simeq 1 - \prod_{l \neq m} \left(1 - \rho_T \rho_S \left\langle \sigma_{i,j}^{(l,m)} \right\rangle \right),$$
$$\tag{5}$$

where in the last term we take advantage of neglecting correlations by replacing the average of the product of variables with the product of the averages and using Eq. (4). This expression simply states that each possible source-target pair weights in the average with the product of the probability that the end nodes are a source and a target; the discovery probability is thus obtained by considering the edge in an average effective media (*mean-field*) of sources and targets homogeneously distributed in the network. This approach is indeed akin to mean-field methods customarily used in the study of many particle systems where each particle is considered in an effective average medium defined by the uncorrelated averages of quantities. The realization average of $\left\langle \sigma_{i,j}^{(l,m)} \right\rangle$ is very simple in the uncorrelated picture, depending only of the kind of $\mathcal{M}$-path. In the case of the ASP probing we have $\left\langle \sigma_{i,j}^{(l,m)} \right\rangle = \sigma_{i,j}^{(l,m)}$, in that each path contributes to the discovery of the edge. In the case of the USP and the RSP, however, only one path among all the equivalent ones is chosen, and in the average we have that each shortest path gives a contribution $\sigma_{i,j}^{(l,m)}/\sigma^{(l,m)}$ to $\left\langle \sigma_{i,j}^{(l,m)} \right\rangle$, where $\sigma^{(l,m)}$ is the number of equivalent shortest path between vertices l and m.

The standard situation we consider is the one in which $\rho_T \rho_S \ll 1$ and since $\left\langle \sigma_{i,j}^{(l,m)} \right\rangle \leq 1$, we have

$$\prod_{l \neq m} \left(1 - \rho_T \rho_S \left\langle \sigma_{i,j}^{(l,m)} \right\rangle \right) \simeq \prod_{l \neq m} \exp \left(-\rho_T \rho_S \left\langle \sigma_{i,j}^{(l,m)} \right\rangle \right), \tag{6}$$

that inserted in Eq.(5) yields

$$\langle \pi_{i,j} \rangle \simeq 1 - \prod_{l \neq m} \left(\exp \left(-\rho_T \rho_S \left\langle \sigma_{i,j}^{(l,m)} \right\rangle \right) \right) = 1 - \exp \left(-\rho_T \rho_S b_{i,j} \right), \qquad (7)$$

where $b_{i,j} = \sum_{l \neq m} \left\langle \sigma_{i,j}^{(l,m)} \right\rangle$. In the case of the USP and RSP probing, the quantity $b_{i,j}$ is by definition the edge betweenness centrality [22, 23], sometimes also referred to as "load" [24] (In the case of ASP probing, it is a closely related quantity). Indeed the vertex or edge betweenness is defined as the total number of shortest paths among pairs of vertices in the network that pass through a vertex or an edge, respectively. If there are multiple shortest paths between a pair of vertices, each path contributes to the betweenness with the corresponding relative weight. The betweenness gives a measure of the amount of all-to-all traffic that goes through an edge or vertex, if the shortest path is used as the metric defining the optimal path between pairs of vertices, and it can be considered as a non-local measure of the *centrality* of an edge or vertex in the graph. It assumes values between 2 and $N(N-1)$.

The discovery probability of the edge will therefore depend strongly on its betweenness. In particular, for vertices with minimum betweenness $b_{i,j} = 2$ we have $\langle \pi_{i,j} \rangle \simeq 2\rho_T \rho_S$, that recovers the probability that the two end vertices of the edge are chosen as source and target. This implies that if the densities of sources and targets are small but finite in the limit of very large N, all the edges have an appreciable probability to be discovered. Moreover, for a large majority of edges with high betweenness the discovery probability approaches one and we can reasonably expect to have a fair sampling of the network.

In most realistic samplings, however, we face a very different situation. While it is reasonable to consider ρ_T a small but finite value, the number of sources is not extensive ($N_S \sim \mathcal{O}(1)$) and their density tends to zero as N^{-1}. In this case it is more convenient to express the edge discovery probability as

$$\langle \pi_{i,j} \rangle \simeq 1 - \exp \left(-\epsilon \overline{b_{i,j}} \right), \qquad (8)$$

where $\epsilon = \rho_T N_S$ is the density of probes imposed to the system and the rescaled betweenness $\overline{b_{i,j}} = N^{-1} b_{i,j}$ is now limited in the interval $[2N^{-1}, N-1]$. In the limit of large networks $N \to \infty$ it is clear that edges with low betweenness have $\langle \pi_{i,j} \rangle \sim \mathcal{O}(N^{-1})$, for any finite value of ϵ. This readily tells us that in real situations the discovery process is generally not complete, a large part of low betweenness edges being not discovered, and that the network sampling is made progressively more accurate by increasing the density of probes ϵ.

A similar analysis can be performed for the discovery indicator function π_i of a vertex i. For each source-target set Ω we have that

$$\pi_i = 1 - \left(1 - \sum_{s=1}^{N_S} \delta_{i,i_s} - \sum_{t=1}^{N_T} \delta_{i,j_t} \right) \prod_{l \neq m \neq i} \left(1 - \sum_{s=1}^{N_S} \delta_{l,i_s} \sum_{t=1}^{N_T} \delta_{m,j_t} \sigma_i^{(l,m)} \right). \qquad (9)$$

where $\sigma_i^{(l,m)} = 1$ if the vertex i belongs to the $\mathcal{M}$-path between nodes l and m, and 0 otherwise. This time it has been considered that each vertex is discovered with probability one also if it is in the set of sources and targets. The second term on the right hand side therefore expresses the probability that the vertex i does not belong to the set of sources and targets and it is not discovered by any $\mathcal{M}$-path between source-target pairs. By using the same *mean-field* approximation as previously, the average vertex discovery probability reads as

$$\langle \pi_i \rangle \simeq 1 - (1 - \rho_S - \rho_T) \prod_{l \neq m \neq i} \left(1 - \rho_T \rho_S \left\langle \sigma_i^{(l,m)} \right\rangle \right). \tag{10}$$

As for the case of the edge discovery probability, the average considers all possible source-target pairs weighted with probability $\rho_T \rho_S$. Also in this case, each shortest path gives a contribution $\sigma_i^{(l,m)} / \sigma^{(l,m)}$ to $\left\langle \sigma_i^{(l,m)} \right\rangle$ for the USP and RSP models, while $\left\langle \sigma_i^{(l,m)} \right\rangle = \sigma_i^{(l,m)}$ for the ASP model. If $\rho_T \rho_S \ll 1$, by using the same approximations used to obtain Eq.(7) we obtain

$$\langle \pi_i \rangle \simeq 1 - (1 - \rho_S - \rho_T) \exp(-\rho_T \rho_S b_i), \tag{11}$$

where $b_i = \sum_{l \neq m \neq i} \left\langle \sigma_i^{(l,m)} \right\rangle$. For the USP and RSP we have that b_i is the vertex betweenness centrality that is limited in the interval $[0, N(N-1)]$ [22, 23, 24]. The betweenness value $b_i = 0$ holds for the leafs of the graph, i.e. vertices with a single edge, for which we recover $\langle \pi_i \rangle \simeq \rho_S + \rho_T$. Indeed, this kind of vertices are dangling ends discovered only if they are either a source or target themselves.

As discussed before, the most usual setup corresponds to a density $\rho_S \sim \mathcal{O}(N^{-1})$ and in the large N limit we can conveniently write

$$\langle \pi_i \rangle \simeq 1 - (1 - \rho_T) \exp\left(-\epsilon \overline{b}_i\right), \tag{12}$$

where we have neglected terms of order $\mathcal{O}(N^{-1})$ and the rescaled betweenness $\overline{b}_i = N^{-1} b_i$ is now defined in the interval $[0, N-1]$. This expression points out that the probability of vertex discovery is favored by the deployment of a finite density of targets that defines its lower bound.

We can also provide a simple approximation for the effective average degree $\langle k_i^* \rangle$ of the node i discovered by our sampling process. Each edge departing from the vertex will contribute proportionally to its discovery probability, yielding

$$\langle k_i^* \rangle = \sum_j \left(1 - \exp\left(-\epsilon \overline{b_{i,j}}\right) \right) \simeq \epsilon \sum_j \overline{b_{i,j}}. \tag{13}$$

The final expression is obtained for edges with $\epsilon \overline{b_{i,j}} \ll 1$. In this case, the sum over all neighbors of the edge betweenness is simply related to the vertex betweenness as $\sum_j b_{i,j} = 2(b_i + N - 1)$, where the factor 2 considers that each vertex path traverses two edges and the term $N - 1$ accounts for all the edge paths for which the vertex is an endpoint. This finally yields

$$\langle k_i^* \rangle \simeq 2\epsilon + 2\epsilon \overline{b}_i. \tag{14}$$

The present analysis shows that the measured quantities and statistical properties of the sampled graph strongly depend on the parameters of the experimental setup and the topology of the underlying graph. The latter dependence is exploited by the key role played by edge and vertex betweenness in the expressions characterizing the graph discovery. The betweenness is a nonlocal topological quantity whose properties change considerably depending on the kind of graph considered. This allows an intuitive understanding of the fact that graphs with diverse topological properties deliver different answer to sampling experiments.

4   Numerical Exploration of Graphs

In the following, we will analyze sparse undirected graphs denoted by $G = (V, E)$. In particular we will consider two main classes of graphs: i) *Homogeneous graphs* in which, for large degree k, the degree distribution $P(k)$ decays exponentially or faster; ii) *Scale-free graphs* for which $P(k)$ has a heavy tail decaying as a power-law $P(k) \sim k^{-\gamma}$. Here the *homogeneity* refers to the existence of a meaningful characteristic average degree that represents the typical value in the graph. Indeed, in graphs with poissonian-like degree distribution a vast majority of vertices has degree close to the average value and deviations from the average are exponentially small in number. On the contrary, scale-free graphs are very heterogeneous with very large fluctuations of the degree, characterized by a variance of the degree distribution which diverges with the size of the network.

Another important characteristic discriminating the topology of graphs is the clustering coefficient that measures the local cohesiveness of nodes. It indeed gives the fraction of connected neighbors of a given node. The average clustering coefficient C provides an indication of the global level of cohesiveness of the graph. The number is generally very small in random graphs that lack of correlations. In many real graphs the clustering coefficient appears to be very high and opportune models have been formulated to represent this property.

4.1   Sampling Homogeneous Graphs

Our first set of experiments considers underlying graphs with homogeneous connectivity; namely the Erdös-Rényi (ER) and the Watts-Strogatz (WS) models. The classical ER model [25] for random graphs $G_{N,p}$, consists of N nodes, each edge being present in E independently with probability p. Since the expected number of edges is $|E| = pN(N-1)/2$, p is of order $1/N$ for sparse graphs. Erdös-Rényi graphs are a typical example of homogeneous graph, with degree distribution following a Poisson law, and clustering coefficient of order $1/N$.

Small-world WS networks are constructed as follows [26]: starting from a one-dimensional lattice of length N, with periodic boundary conditions (i.e. a ring), each vertex being connected to its $2m$ nearest neighbors (with $m > 1$). The vertices are then visited one after the other; each link connecting a vertex to one of its m nearest neighbors in the clockwise sense is left in place with probability $1 - p$, and with probability p is reconnected to a randomly chosen

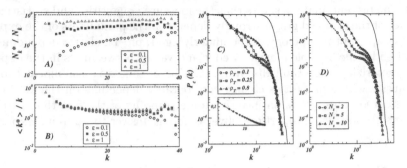

Fig. 1. A) Frequency N_k^*/N_k of detecting a vertex of degree k; B) proportion of discovered edges $\langle k^* \rangle /k$ as a function of the degree in the ER graphs. The exploration setup considers $N_S = 2$ and increasing probing level $\epsilon = N_S \rho_T$ obtained by progressively higher density of targets ρ_T. The y axis is in log scale to allow a finer resolution. C) and D) Cumulative degree distribution of the sampled ER graph for USP probes. Figure C) shows sampled distributions obtained with $N_S = 2$ and varying density target ρ_T. Inset: case $N_S = 1$, providing an apparent power-law behavior with exponent -1 at all values of ρ_T. The inset is in lin-log scale to show the logarithmic behavior of the corresponding cumulative distribution. The sampled distributions of figure D) are obtained with $\rho_T = 0.1$ and varying number of sources N_S. The solid line is the degree distribution of the underlying graph

other vertex, introducing therefore shortcuts. The number of edges is $|E| = Nm$, independently of p. The degree distribution has a shape similar to the case of ER graphs, peaked around the average value. The clustering coefficient, however, is large if $p \ll 1$, making this network a typical example of homogeneous but clustered network. In both ER and WS cases, graphs may consist of more than one connected component; the largest of these components is then used. We have used networks of $N = 10^4$ nodes, $\overline{k} = 20$ unless otherwise specified; for the WS model, $p = 0.1$ has been taken. Each measure is averaged over 10 realizations.

Analogously to the degree distribution, the vertex and edge betweenness distributions are narrowly distributed around the average values $\overline{b}$ and $\overline{b}_e$, respectively, which can thus be considered as typical values. Since a large majority of vertices and edges will have a betweenness very close to the average value, we can use Eq. (8) and (12) to estimate the order of magnitude of probes that allows a fair sampling of the graph. Indeed, both $\langle \pi_{i,j} \rangle$ and $\langle \pi_i \rangle$ tend to 1 if $\epsilon \gg \max \left[\overline{b}^{-1}, \overline{b}_e^{-1} \right]$. In this limit all edges and vertices will have probability to be discovered very close to one. At lower ϵ, obtained by varying ρ_T and N_S, the underlying graph is only partially discovered. We first report in Fig. 1, for both the ER and WS models, the behavior of the fraction N_k^*/N_k of discovered vertices of degree k, where N_k is the total number of vertices of degree k in the underlying graph, and the fraction of discovered edges $\langle k^* \rangle /k$ in vertices of degree k. The fraction N_k^*/N_k naturally increases by augmenting the density of targets and sources, and it is slightly increasing for larger degrees. The latter behavior can be easily understood by noticing that vertices with larger degree have on average a larger betweenness $b(k)$. By using Eq.(12) we have that $N_k^*/N_k \sim 1 - \exp\left(-\epsilon\overline{b(k)}\right)$, obtaining the observed increase at large k. On the other hand, the range of variation of degree in homogeneous graph is very narrow and only a large level of probing may guarantee very large discovery probabilities. Similarly the behavior of the effective discovered degree can be understood by looking at Eq. (14) stating that $\langle k^* \rangle / k \simeq \epsilon k^{-1}(1 + \overline{b(k)})$. Indeed the initial decrease of $\langle k^* \rangle / k$ is finally compensated by the increase of $\overline{b(k)}$.

A very important quantity in the study of the statistical accuracy of the sampled graph is the degree distribution. In Fig. 1 we show the cumulative degree distribution $P_c(k^* > k)$ of the sampled graph defined by the ER model for increasing density of targets and sources. Sampled distributions are only approximating the genuine distribution, however, for $N_S \geq 2$ they are far from a true heavy-tail distributions at any appreciable level of probing. Indeed, the distribution runs generally over a small range of degrees, with a cut-off that sets in at the average degree $\overline{k}$ of the underlying graph. In order to stretch the distribution range, homogeneous graphs with very large average degree $\overline{k}$ must be considered; however, other distinctive spurious effects appear in this case. In particular, since the best sampling occurs around the high degree values, the distributions develop peaks that show in the cumulative distribution as plateaus. The very same behavior is obtained in the case of the WS model. Finally, in the case of RSP and ASP model, we observe that the obtained distributions are closer to the real one since they allow a larger number of discoveries.

Only in the peculiar case of $N_S = 1$ an apparent scale-free behavior with slope -1 is observed for all target densities ρ_T, as analytically shown by Clauset and Moore [20]. Also in this case, the distribution cut-off is consistently determined by the average degree $\overline{k}$. It is worth noting that the experimental setup with a single source is a limit case corresponding to a highly asymmetric probing process; it is therefore badly, if at all, captured by our statistical analysis which assumes homogeneous deployment.

The present analysis shows that in order to obtain a sampled graph with apparent scale-free behavior on a degree range varying over n orders of magnitude we would need the very peculiar sampling of a homogeneous underlying graph with an average degree $\overline{k} \simeq 10^n$; a rather unrealistic situation in the Internet and many other information systems where $n \geq 2$.

4.2   Sampling Scale-Free Graphs

In this section, we extend the analysis made for homogeneous graphs to the case of highly heterogeneous scale-free graphs.

Albert and Barabási (BA) have proposed to combine two ingredients to obtain a heterogeneous scale-free graph [27]: i) *Growth:* Starting from an initial seed of N_0 vertices connected by links, a new vertex n is added at each time step; ii) This new site is connected to m previously existing vertices chosen with the *preferential attachment rule:* a node i is chosen by n with a probability pro-

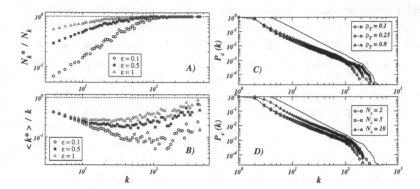

Fig. 2. A) Frequency N_k^*/N_k of detecting a vertex of degree k; B) proportion of discovered edges $\langle k^* \rangle /k$ (bottom) as a function of the degree in the BA graphs. The exploration setup considers $N_S = 2$ and increasing probing level ϵ obtained by progressively higher density of targets ρ_T. The plot is in log-log scale to allow a finer resolution and account for the wide variation of degree in scale-free graphs. C) and D) Cumulative degree distribution of the sampled BA graph for USP probes. Figure C) shows sampled distributions obtained with $N_S = 2$ and varying density target ρ_T. The data on figure D) are obtained with $\rho_T = 0.1$ and varying number of sources N_S. The solid line is the degree distribution of the underlying graph

portional to its degree. The growth process is then iterated by introducing a new vertex, i.e. going back to step i) until the desired size of the network is reached. This mechanism yields a connected graph of $|V| = N$ nodes with $|E| = mN$ edges, with a power-law degree distribution $P(k) \sim k^{-\gamma}$ with $\gamma = 3$, and a small clustering coefficient.

Dorogovtsev, Mendes and Samukhin (DMS) have introduced [28] a model of growing network with very large clustering coefficient C: at each time step, a new node is introduced and connected to *the two extremities of a randomly chosen edge*, thus forming a triangle. A given node is thus chosen with a probability proportional to its degree, which corresponds to the preferential attachment rule. The graphs obtained have N nodes, $2N$ edges, a large clustering coefficient (≈ 0.74) and a power-law distribution for the degree distribution of the nodes.

We have used networks of size $N = 10^4$ ($\bar{k} = 8$ for BA and $\bar{k} = 4$ for DMS), and averaged each measurement over 10 realizations. The average degree of both models is well defined, however, the degree distribution is heavy-tailed with fluctuations diverging logarithmically with the graph size. This implies that $\bar{k}$ is not a typical value in the network and there is an appreciable probability of finding vertices with very high degree. Analogously, the betweenness distribution is heavy-tailed, allowing for an appreciable fraction of vertices and edges with very high betweenness [29]. In fact it is possible to show that in scale-free graphs the site betweenness is related to the degree as $\overline{b(k)} \sim k^\beta$, where β is an exponent depending on the model [29]. Since in heavy-tailed distributions the degree is varying over several orders of magnitude, the same occurs for the betweenness values. In such a situation, even in the case of small ϵ, vertices whose betweenness is large enough $(\overline{b(k)}\epsilon \gg 1)$ have $\langle \pi_i \rangle \simeq 1$. Therefore all vertices with degree $k \gg \epsilon^{-1/\beta}$ will be detected with probability one. Fig. 2 indeed clearly shows that the discovery probability N_k^*/N_k of vertices with degree k saturates to one for large degree values, at a value of k which decreases with increasing ϵ. The measures of $\langle k^* \rangle / k$ show a similar effect. After an initial decay (see Fig. 2) the effective discovered degree increases with the degree of the vertices, as predicted by Eq. (14) that gives $\langle k^* \rangle / k \simeq \epsilon k^{-1}(1 + \overline{b(k)})$: at large k the term $k^{-1}\overline{b(k)} \sim k^{\beta-1}$ takes over and the effective discovered degree approaches the real degree k.

It is evident from the previous discussions, that in scale-free graphs, vertices with high degree are efficiently sampled with an effective measured degree that is rather close to the real one. This means that the degree distribution tail is fairly well sampled while deviations should be expected at lower degree values. This is indeed what we observe in numerical experiments on BA and DMS graphs (see Fig 2). Despite both underlying graphs have a small average degree, the observed degree distribution spans more than two orders of magnitude. The distribution tail is fairly reproduced even at rather small values of ϵ, while the low degree regime is instead under-sampled providing an apparent change in the exponent of the degree distribution. This effect has been noticed also by Petermann and De Los Rios in the case of single source experiments [21].

The present analysis points out that graphs with heavy-tailed degree distribution allow a better qualitative representation of their statistical features in sampling experiments. Indeed, the most important properties of these graphs are related to the heavy-tail part of the statistical distributions that are indeed well discriminated by the `traceroute`-like exploration.

5   Conclusions and Outlook

The analysis we have presented allows the understanding of the effect of the underlying graph topology in the sampling process. Indeed, the rationalization of the exploration biases at the statistical level provides a general interpretative framework for the results obtained from the numerical experiments on graph models. The exploration process, which focuses on high betweenness nodes, provides a very accurate sampling of the distribution tail, and thus allows to clearly distinguish the two situations defined by homogeneous and heavy-tailed topologies. In graphs with heavy-tails, such as scale-free networks, the main topological features are easily discriminated since the relevant statistical information is encapsulated in the degree distribution tail which is fairly well captured. Quite surprisingly, the sampling of homogeneous graphs appears more cumbersome than those of heavy-tailed graphs. Dramatic effects such as the existence of apparent power-laws, however, are found only in very peculiar cases. In general, exploration strategies provide sampled distributions with enough signatures to distinguish at the statistical level between graphs with different topologies.

This evidence might be relevant in the discussion of real data from Internet mapping projects, which indicate the presence of heavy-tailed degree distribution both at the router and AS level. In the light of the present discussion, it is very unlikely that this feature is just an artifact of the mapping strategies. The upper degree cut-off at the router and AS level runs up to 10^2 and 10^3, respectively. A homogeneous graph should have an average degree comparable to the measured cut-off, hardly conceivable in a realistic perspective (it would e.g. require that nine routers over ten would have more than 100 links to other routers). Moreover, the major part of mapping projects are multi-source, a feature that we have shown to readily wash out the presence of spurious heavy-tailed behavior. On the contrary, heavy tails are easily sampled with particular accuracy for the large degree part, generally at all probing levels. This makes very plausible, and a natural consequence, that the heavy-tail behavior observed in real mapping experiments is a genuine feature of the Internet.

On the other hand, it is important to stress that while at the qualitative level the sampled graphs allow a good discrimination of the statistical properties, at the quantitative level they might exhibit considerable deviations from the true values such as average degree, distribution exponent and clustering properties. In this respect, it is of major importance to define strategies that optimize the estimate of the various parameters and quantities of the underlying graph. For instance, the deployment of a highly distributed infrastructure of sources probing a limited number of targets may result as efficient as a few very powerful sources probing a large fraction of the addressable space. The optimization of large network sampling is therefore an open problem that calls for further work aimed at a more quantitative assessment of the mapping strategies both on the analytic and numerical side.

Acknowledgments. This work has been partially supported by the European Commission Fet-Open project COSIN IST-2001-33555 and contract 001907 (DELIS).

References

1. The National Laboratory for Applied Network Research (NLANR), sponsored by the National Science Foundation. (see http://moat.nlanr.net/).
2. The Cooperative Association for Internet Data Analysis (CAIDA), located at the San Diego Supercomputer Center. (see http://www.caida.org/home/).
3. Topology project, Electric Engineering and Computer Science Department, University of Michigan (http://topology.eecs.umich.edu/).
4. SCAN project, Information Sciences Institute (http://www.isi.edu/div7/scan/).
5. Internet mapping project at Lucent Bell Labs (http://www.cs.bell-labs.com/who/ches/map/).
6. M. Faloutsos, P. Faloutsos, and C. Faloutsos, "On Power-law Relationships of the Internet Topology," ACM SIGCOMM '99, Comput. Commun. Rev. **29**, 251–262 (1999).
7. R. Govindan and H. Tangmunarunkit, "Heuristics for Internet Map Discovery," Proc. of IEEE Infocom 2000, Volume 3, 1371–1380, (2000).

8. A. Broido and K. C. Claffy, "Internet topology: connectivity of IP graphs," San Diego Proceedings of SPIE International symposium on Convergence of IT and Communication. Denver, CO. 2001

9. G. Caldarelli, R. Marchetti, and L. Pietronero, "The Fractal Properties of Internet," Europhys. Lett. **52**, 386 (2000).

10. R. Pastor-Satorras, A. Vázquez, and A. Vespignani, "Dynamical and Correlation Properties of the Internet," Phys. Rev. Lett. **87**, 258701 (2001); A. Vázquez, R. Pastor-Satorras, and A. Vespignani, "Large-scale topological and dynamical properties of the Internet," Phys. Rev. E .**65**, 066130 (2002).

11. Q. Chen, H. Chang, R. Govindan, S. Jamin, S. J. Shenker, and W. Willinger, "The Origin of Power Laws in Internet Topologies Revisited," Proceedings of IEEE Infocom 2002, New York, USA.

12. A. Medina and I. Matta, "BRITE: a flexible generator of Internet topologies," Tech. Rep. BU-CS-TR-2000-005, Boston University, 2000.

13. C. Jin, Q. Chen, and S. Jamin, "INET: Internet topology generators," Tech. Rep. CSE-TR-433-00, EECS Dept., University of Michigan, 2000.

14. S. N. Dorogovtsev and J. F. F. Mendes, *Evolution of networks: From biological nets to the Internet and WWW* (Oxford University Press, Oxford, 2003).

15. P.Baldi, P.Frasconi and P.Smyth, *Modeling the Internet and the Web: Probabilistic methods and algorithms*(Wiley, Chichester, 2003).

16. R. Pastor-Satorras and A. Vespignani, *Evolution and structure of the Internet: A statistical physics approach* (Cambridge University Press, Cambridge, 2004).

17. H. Burch and B. Cheswick, "Mapping the internet," IEEE computer, **32(4)**, 97–98 (1999).

18. W. Willinger, R. Govindan, S. Jamin, V. Paxson, and S. Shenker, "Scaling phenomena in the Internet: Critically examining criticality," Proc. Natl. Acad. Sci USA **99** 2573–2580, (2002).

19. A. Lakhina, J. W. Byers, M. Crovella and P. Xie, "Sampling Biases in IP Topology Measurements," Technical Report BUCS-TR-2002-021, Department of Computer Sciences, Boston University (2002).

20. A. Clauset and C. Moore, "Traceroute sampling makes random graphs appear to have power law degree distributions," arXiv:cond-mat/0312674 (2003).

21. T. Petermann and P. De Los Rios, "Exploration of scale-free networks - Do we measure the real exponents?", Eur. Phys. J B **38**, 201 (2004).

22. L. C. Freeman, "A Set of Measures of Centrality Based on Betweenness," Sociometry **40**, 35–41 (1977).

23. U. Brandes, "A Faster Algorithm for Betweenness Centrality," J. Math. Soc. **25(2)**, 163–177, (2001).

24. K.-I. Goh, B. Kahng, and D. Kim, "Universal Behavior of Load Distribution in Scale-Free Networks," Phys. Rev. Lett. **87**, 278701 (2001).

25. P. Erdös and P. Rényi, "On random graphs I," Publ. Math. Inst. Hung. Acad. Sci. **5**, 17 (1960).

26. D. J. Watts and S. H. Strogatz, "Collective dynamics of small-world networks," Nature **393**, 440–442 (1998).

27. A.-L. Barabási and R. Albert, "Emergence of scaling in random networks," Science **286**, 509–512 (1999).

28. S. N. Dorogovtsev, J. F. F. Mendes, and A. N. Samukhin, "Size-dependent degree distribution of a scale-free growing network," Phys. Rev. E **63**, 062101 (2001)

29. M. Barthélemy, "Betweenness Centrality in Large Complex Networks," Eur. Phys. J B **38**, 163 (2004).

Invited Talk: The Many Wonders
of the Web Graph

Andrei Broder

Distinguished Engineer, IBM Research

Abstract. The Web graph, meaning the graph induced by Web pages as nodes and their hyperlinks as directed edges, has become a fascinating object of study for many people: physicists, sociologists, mathematicians, computer scientists, and information retrieval specialists.

Recent results range from theoretical (e.g.: models for the graph, semi-external algorithms), to experimental (e.g.: new insights regarding the rate of change of pages, new data on the distribution of degrees), to practical (e.g.: improvements in crawling technology, uses in information retrieval, web spam prevention).

The goal of this talk is to convey an introduction to the state of the art in this area and to sketch the current issues in collecting, representing, analyzing, and modeling this graph.

A. López-Ortiz and A. Hamel (Eds.): CAAN 2004, LNCS 3405, p. 154, 2005.
© Springer-Verlag Berlin Heidelberg 2005

Algorithmic Foundations of the Internet:*

Foreword

Alejandro López-Ortiz

School of Computer Science, University of Waterloo,
Waterloo, ON, Canada
alopez-o@uwaterloo.ca

1 Introduction

During the last 25 years the Internet has grown from being a small academic network connecting a few computer science departments to its present size, connecting more than 285 million computers and serving over 800 million users worldwide [7].

With the introduction of the Mosaic web browser in 1993, the rate of growth accelerated both in terms of traffic and in number of users. It became clear then that the Internet would be large enough to make small performance differences significant. For example, for many realistic problem sizes outside the Internet, the difference between a $\Theta(n \log n)$ and a $\Theta(n)$ algorithm is not terribly significant, since $\log n$ is a such a slowly growing function. This is particularly so if the $\Theta(n \log n)$ solution has a simpler implementation resulting in a smaller leading constant. In contrast $\log_2 n$ on the Internet is most often larger than 20 and in some instances as large as 50. In such cases, for example, a solution requiring n bits of storage is twenty or possibly fifty times smaller than a solution requiring space $n \log n$. This is by all means a non-negligible factor.

Algorithmic foundations of the Internet is a new field within theoretical computer science designed to address the challenges posed by such large scale networks. It lies in the intersection of networks, graph theory, algorithms and game theory. Other areas of relevance are distributed computing, statistical physics, economics, information retrieval and probability theory. Its early beginnings coincide with the popularization of the world wide web—and hence the Internet—during 1993-1995, and has been a rapidly expanding area of research ever since.

This field is, in a sense, a "theory of internetworking". This new theory of networking brings to bear tools that are of common use, so to speak, within the field of algorithms and theoretical computer science and applies them to networking problems. In particular it focuses on problems posed by the current Internet given its size and the specific set of protocols defining it. Early efforts in this field, as we shall see, were mostly algorithmic, e.g. how to compute bet-

---

* An earlier version of this survey appearead in "Algorithmic Foundations of the Internet", A. López-Ortiz, SIGACT News, v. 36, no. 2, June 2005.

A. López-Ortiz and A. Hamel (Eds.): CAAN 2004, LNCS 3405, pp. 155–158, 2005.

ter routing algorithms, shortest path distributed computations, better indexing algorithms for search engines, etc. Yet, as the field has grown it has incorporated techniques from graph theory, economics and game theory, statistics and probability, as well as statistical physics.

As well the field has received ample coverage from the popular press. This is in part due to the elegance of some of the results and in part to the fact that results can often be explained to the general public in approachable terms. For an enjoyable introduction some of the topics in this survey, see for example [1, 3, 2, 4, 5, 6, 8].

2   Historical Perspective

During the first fifteen years of the Internet, its initial design generally proved highly scalable and its growth was accommodated organically. Only in the early 1980's—a full decade after the network had first gone live in a form reasonably approaching the current Internet—did the first hints appear of the need to redesign certain aspects to accommodate impending growth. In particular the method for translating (or resolving) a computer name into an IP address started to show strains due to rapid network growth. Hence in 1984, the Domain Name System (DNS) was introduced. This system translates addresses such as cs.uwaterloo.ca to the numeric address 129.97.152.134. When the DNS system was activated in 1984 it was one of earliest, largest and highest-transaction-rate distributed databases yet. It would take many years before any other in-production distributed database approached the size and transaction-rate of DNS.

Similarly, routing, which is the method used to direct a message across the network, showed strains in the late 1980's. Until then each computer in the network had something approaching a global map of the network in local storage. With the number of hosts rapidly growing such a method became unfeasible. Hence the Border Gateway Protocol (BGP) was introduced in 1989. The BGP protocol changed the routing decision process from a centralized global routing information model (EGP), to an innovative, decentralized protocol in which a node in the network is only responsible for knowing the local routing choice. This is equivalent to setting out to cross the entire continent, asking for directions at a service station and being told only what to do for the next five miles. The advantage is that each service station needs to know only their local neighbourhood; the disadvantage is that it leads to repeated queries throughout the route to the destination. BGP introduced various innovations which were extremely well suited for theoretical analysis, and indeed some analysis took place early on, e.g. on the number of messages needed to disseminate routes starting from a blank network. Yet only a full five years after the BGP4 version of the protocol was introduced, did theoretical analysis show that, in the worst case, it could take as many as $\Theta(n!)$ steps to fully disseminate and stabilize a change on a network of n nodes [9].

The popularization of the World Wide Web in 1993 introduced yet another set of large scale challenges. For example, by 1995, the most advanced text indexing algorithms to that point had yet to meet their match within corporate applications. In contrast, by the end of 1995, those same algorithms were straining to sustain normal search engine loads on the Web. The situation has not improved much since, as all algorithmic advances in that field are quickly subsumed by the rapid growth in the web.

A second example is given by routing algorithms. The first routing algorithm devised had time complexity of $\Theta(\log n)$, which is generally considered fast as far as algorithms are concerned, yet today routers are strained using an improved $\Theta(\log \log n)$ algorithm. Again in most other settings the difference between those two time complexities would be too small to be of any impact. Yet on the Internet it is crucially significant.

3   Structure of the Surveys

We present three long surveys as well as a short a roundup of three other areas. The long surveys are:

- an introduction to models for the web graph by Anthony Bonato,
- an overview on routing by Angele Hamel and Jean-Louis Gregoire, and
- a survey on search engines and web information retrieval by Alejandro López-Ortiz

The roundup gives a short introduction to

- web caching,
- internet economics and game theory, and
- tomography.

This selection of topics is by no means exhaustive, rather it was chosen as illustrative of the variety of algorithmic challenges arising from the Internet.

References

1. R. Albert, H. Jeong, and A.-L. Barabśi. Error and attack tolerance of complex networks. Nature 406, 378-482 (2000).
2. A.-L. Barabási. The physics of the Web. *Physics World*, 14, 33 (2001).
3. A.-L. Barabási, R. Albert, H. Jeong, and G. Bianconi. Power-Law Distribution of the World Wide Web, Science 287, 2115 (2000).
4. A.-L. Barabási, and E. Bonabeau. Scale-free networks. *Scientific American*, 288(5):60, 2003.
5. S. Chakrabarti, B. Dom, D. Gibson, J. Kleinberg, S.R. Kumar, P. Raghavan, S. Rajagopalan, and A. Tomkins. Hypersearching the Web. *Scientific American*, June 1999.
6. S. Chakrabarti, B. Dom, D. Gibson, J. Kleinberg, S.R. Kumar, P. Raghavan, S. Rajagopalan, and A. Tomkins. Mining the link structure of the World Wide Web. *IEEE Computer*, August 1999.

7. Internet Software Consortium http://www.isc.org.
8. H. Jeong, B. Tombor, R. Albert, Z. Oltvai, and A.-L. Barabsi. The large-scale organization of metabolic networks. *Nature* 407, pp. 651–654, 2000.
9. Craig Labovitz, Abha Ahuja, Abhijit Bose, and Farnam Jahanian. Delayed internet routing convergence. In Proceedings of *ACM Conference on Applications, Technologies Architectures and Protocols for Computer Communication* (SIGCOMM) 2000, pp. 175–187.

A Survey of Models of the Web Graph

Anthony Bonato*

Department of Mathematics,
Wilfrid Laurier University,
Waterloo, ON, Canada, N2L 3C5
abonato@rogers.com

Abstract. The web graph has been the focus of much recent attention, with several stochastic models proposed to account for its various properties. A survey of these models is presented, focussing on the models which have been defined and analyzed rigorously.

1 Introduction

The web graph, which we will denote by W, has nodes representing web pages, and edges representing the links between pages. The graph W is massive: at the time of writing, it contains billions of nodes and edges. In addition, W is dynamic or *evolving*, with nodes and edges appearing and disappearing over time. The explosive growth of W itself is mirrored by the recent rapid increase in research on structural properties, stochastic models, and mining of the web graph. There are now several survey articles [10, 19, 44], and several books on W [6, 22, 31] (including a popular book by Barabási [7]). A new mathematics journal devoted to research related to W, *Internet Mathematics*, was recently launched.

The purpose of the present survey is to highlight recent stochastic models which are used to model W. We focus on six of these models, chosen both for their various design elements, and because they have been rigorously defined and analyzed. For more on information on web mining and the mathematics of web search engines, the reader is directed to Chakrabarti [22] and Henzinger [39].

For background on graph theory and random graphs, the reader is directed to [8, 9, 30, 40]. We use the notation $\mathbb{N}$ for the nonnegative integers and $\mathbb{N}^+$ for the positive integers. If A is an event in a probability space Ω, then $\mathbb{P}(A)$ is the probability of the event in the space Ω; if X is a random variable with domain Ω, then $\mathbb{E}(X)$ is the expectation of X. All logarithms are in base 2.

2 Properties of W

We begin with a brief overview of some of the experimental data and structural properties observed in W collected from various web crawls. The overview will high-

---

* The author gratefully acknowledges the support from an NSERC Discovery grant and from a MITACS grant.

A. López-Ortiz and A. Hamel (Eds.): CAAN 2004, LNCS 3405, pp. 159–172, 2005.

light experimental features of W that models of the web graph often attempt to replicate. For a more detailed introduction to experimental data on W, the reader is directed to Chapter 3 of Dorogovtsev, Mendes [31] and Kumar et al. [44].

Arguably the most important properties observed in W are power-law degree distributions. Given an undirected graph G and a non-negative integer k, we define the rational number $P_G(k)$ as follows.

$$P_G(k) = \frac{|\{x \in V(G) : \deg_G(x) = k\}|}{|V(G)|}.$$

In other words, $P_G(k)$ is the proportion of nodes of degree k in G. We will suppress the subscript G if it is clear from context. We say that the degree distribution of G follows a *power law* if for each degree k,

$$P(k) \sim ck^{-\beta},$$

for real constants $c > 0$ and $\beta > 0$. Such distributions are sometimes called *heavy-tailed distributions*, since the real-valued function $f(k) = ck^{-\beta}$ exhibits a polynomial decay to 0 as k tends to ∞. Real-world graphs like W with power law degree distributions are sometimes called *scale-free*. The graph W may be viewed as either a directed or undirected graph. If G is directed, then we may discuss power laws for the in- and out-degree distributions by defining the proportions $P_{in,G}(k)$ and $P_{out,G}(k)$, respectively, in the obvious way.

Based on their crawl of the domain of Notre Dame University, Indiana, Albert et al. [4] claimed that the web graph exhibits a power law in-degree distribution, with $\beta = 2.1$. This claim was supported by an independent larger crawl of the entire web reported in Broder et al. [17], who also found $\beta = 2.1$. There is some evidence in both studies that the out-degree distribution follows a power law with [4] reporting $\beta = 2.45$ and [17] reporting $\beta = 2.7$. The presence of power law degree distributions reflects a certain *undemocratic* aspect of W: while most pages have few links, a few have a large number. This is perhaps not surprising, since the choice of links from new pages to existing ones is presumably governed by the users own personal or commercial interests. It is interesting to note that power law degree distributions are now known to be pervasive in a variety of real world networks where some degree of choice is involved, such as the telephone call network, the e-mail network, or the scientific citation network; see [31] for other similar networks. Power law degree distributions are also prevalent in biological networks (such as the network of protein-protein interactions in a cell), where evolution is the dominant decision making force in the generation of nodes and edges; see Chung et al. [23]. Models for power law behaviour have long been studied in such disciplines as biology and economics; see Mitzenmacher [45].

So-called *small world graphs* were first introduced by Strogatz, Watts [46] in their study of social networks. One important feature of small world networks is the presence of "short" paths between nodes. To be more precise, define the *distance* from u to v in a graph G, written $d(u,v)$, to be the number of edges in a shortest path connecting u to v, or ∞ otherwise. Define

$$L(G) = \sum_{\{u,v\}\in S} \frac{d(u,v)}{|S|},$$

where S is the set of pairs of distinct nodes u, v of G with the property that $d(u,v)$ is finite. The rational number $L(G)$ is the *average distance* of G. The directed analogue of this parameter, where distance refers to shortest *directed* paths, is denoted $L_d(G)$. The small world property demands that $L(G)$ (or $L_d(G)$ if G is directed) must be much smaller than the order of the graph; for example, $L(G) \in O(\log(|V(G)|))$. As evidence of the small world property for W, in Albert et al. [4] it was reported that $L_d(W) = 19$, while Broder et al. [17] reported $L_d(W) = 16$ and $L(W) = 6.8$. Another measure of global distances in a graph is the *diameter* of G, written $diam(G)$, which is the maximum of $d(u,v)$ taken over all pairs of distinct nodes u and v in G. In contrast with the just cited results on average distance, data from [17] suggests that $diam(W) > 900$.

The web contains many *communities*: sets of pages sharing a common interest. A notion presented in Kleinberg et al. [41] and Kumar et al. [44] is that communities in the web are characterized by dense directed bipartite subgraphs. A *bipartite core* is a directed graph which contains at least one directed bipartite clique as a subgraph, where the directed edges in the subgraph all have terminal nodes of one fixed colour. In their study of communities in W, the authors in [41, 44] show the presence of many more small bipartite cores in W than a directed random graph with the same number of nodes and edges.

3   Models of W

A large number of models for the web graph have been proposed. Such models are useful for several reasons. They deepen our understanding of the generative mechanisms driving the evolution of W. They provide insight into superficially unrelated properties observed in the web. Perhaps most importantly from the point of view of applications, they may aid in the development of the next generation of link-analytic search engines. As discussed in the survey by Bollobás, Riordan [10], the majority of the analysis of models of the web has been heuristic and non-rigorous. A small but growing number of rigorous studies of web graph models have been appearing in the literature, and it is these models that we focus on in the present survey.

Pioneering work on random graphs was first done by Erdős and Rényi [34, 35]. In what is now sometimes called the Erdős-Rényi (ER) model, written $G(n,p)$, we are given n nodes and a fixed real number $p \in (0,1)$. For each of the $\binom{n}{2}$ many distinct pairs of nodes, add an edge between them independently with probability p. In many contexts, p is a function of n, and properties of $G(n,p)$ are studied asymptotically as n tends to ∞. The probability space $G(n,p)$ is often referred to as *a random graph of order n* (an accepted misnomer). Random graphs have been intensively researched, and the subject has spawned several thousand research articles. We direct the interested reader to the texts of Bollobás [9] and Janson et al. [40] for more on the ER model.

The ER model is, in a sense, static or *off-line*: the number of nodes is fixed (although $G(n, p)$ is often viewed as having a variable number of edges with time). In addition, it is straightforward to prove that in $G(n, p)$ the degree of a node is binomially distributed. Hence, based on our discussion in Section 1, the ER model is not appropriate as a model of the web graph W. What features would make a good web graph model? The following is a (partial) list of desirable properties that graphs generated by a web graph model should possess, based on the observed properties of W given in the Introduction.

1. *On-line property.* The number of nodes and edges changes with time.
2. *Power law degree distribution.* The degree distribution follows a power law, with an exponent $\beta > 2$.
3. *Small world property.* The average distance (or diameter) is much smaller than the order of the graph.
4. *Many dense bipartite subgraphs.* The number of distinct bipartite cliques or cores is large when compared to a random graph with the same number of nodes and edges.

To aid the reader, we give a chart that summarizes the properties of the various models we will consider. We will focus on the following web graph models, each given an acronym (if they do not already have one) for purposes of comparison: the LCD model of Bollobás et al [14]; the ACL models of Aiello et al [3], the CL model of Chung, Lu [24], [25], the copying model of Kumar et al [43]; the CL-del growth-deletion model of Chung, Lu [27]; and the CFV growth-deletion model of Cooper et al. [29]. A "Y" in the i, j entry of the table means that the model in row i has the property of column j; a "N" or "?" are read similarly. The column "Directed?" refers to whether the model generates directed graphs. The column entitled "β" refers to the possible range of exponent for the power law proven asymptotically for graphs generated by the model, with the value of β dependent on the parameters of the model. If the model produces directed graphs, then the range refers to the in-degree distribution.

Model	Directed?	1	2	3	4	β
LCD	Y	Y	Y	Y	?	3
ACL	Y	Y	Y	?	N	$(2, \infty)$
CL	N	N	Y	Y	?	$(2, \infty)$
copying	Y	Y	Y	?	Y	$(2, \infty)$
CL-del	N	Y	Y	Y	?	$(2, \infty)$
CFV	N	Y	Y	?	?	$(2, \infty)$

As is evident from the chart, all the models in the present survey have property 2, all but the CL model have property 1, while it has yet to be proven that any of them have all four properties. At the time of writing, it is an open problem to find a web graph model that produces graphs which provably has all four properties.

To simplify notation, we present the following general mathematical framework for all the on-line models presented. (Hence, the CL model will not follow this framework.) The model possesses a set of real number parameters, and has a fixed finite graph H as an additional parameter. The model generates by some stochastic process a sequence of finite graphs G_t indexed by $(t : t \in \mathbb{N})$. Unless otherwise stated, for all $t \in \mathbb{N}$, we have that

1. $G_0 \cong H$;
2. G_t is an induced subgraph of G_{t+1};
3. $|V(G_{t+1})| = |V(G_t)| + 1$;
4. $|E(G_t)| \leq |E(G_{t+1})|$.

In all the models we consider, the graphs G_t are defined inductively. In the inductive step, the unique node in $V(G_{t+1})\backslash V(G_t)$ is referred to as the *new node*, written v_{t+1}, and the nodes of $V(G_t)$ are the *existing nodes*. We refer to a model which generates graphs satisfying all of these conditions as an *evolving graph model*. We note that the choice of H usually has no effect on the value of the power law exponent β, while the choice of real number parameters does generally affect β.

4   Preferential Attachment Models

The first evolving graph model explicitly designed to model W was given by Albert, Barabási [5]. Informally, the idea behind their model is a straightforward and intuitively pleasing one: new nodes are more likely to join to existing nodes with high degree. This model is now referred to as an example of a *preferential attachment model*. Albert and Barabási gave a heuristic description and analysis of such a model, and concluded that it generates graphs whose in-degree distribution follows a power law with exponent $\beta = 3$.

The first rigorous attempt to design and analyze a preferential attachment model was given in Bollobás et al. [14]. Their model is called the *Linearized Chord Diagram* or *LCD* model, since an equivalent formulation of the model is via random pairings on a fixed finite set of integers. The parameter of this model is a positive integer m, where H is a copy of K_1 with a single directed loop. We first describe the model in the case $m = 1$. To form G_{t+1} add a single directed edge from v_{t+1} to v_i, where the node v_i is chosen at random from the existing nodes, with

$$\mathbb{P}(i = s) = \begin{cases} \frac{\deg_{G_{t-1}}(v_s)}{2t-1} & \text{if } 1 \leq s \leq t-1, \\ \frac{1}{2t-1} & \text{if } s = t. \end{cases}$$

This mechanism of joining new nodes to existing ones proportionally by degree we refer to as *preferential attachment*. Observe that the graph G_t is a directed tree for all values of t. Indeed a similar version of this model was previously studied (in a different context) as *random recursive trees*; see [14] for further discussion.

If $m > 1$, then define the process $(G_m^t)_{t\geq 0}$ by first generating a sequence $(G_t : t \in \mathbb{N})$ of graphs using the case $m = 1$ on a sequence of nodes $(v_i' : i \in \mathbb{N}^+)$. The graph G_m^t is formed from G_{mt} by identifying the nodes $v_1', v_2', \ldots, v_m'$ to form v_1, identifying $v_{m+1}', v_{m+2}', \ldots, v_{2m}'$ to form v_2, and so on.

Using martingales and the Azuma-Hoeffding inequality (see Theorem 1.19 of [9], for example), Bollobás et al. [14] prove the following theorem.

Theorem 1. *Fix m a positive integer, and fix $\epsilon > 0$. For k a non-negative integer, define*

$$\alpha_{m,k} = \frac{2m(m+1)}{(k+m)(k+m+1)(k+m+2)}.$$

Then with probability tending to 1 as $t \to \infty$, for all k satisfying $0 \leq k \leq t^{1/15}$,

$$(1-\epsilon)\alpha_{m,k} \leq P_{in,G_m^t}(k) \leq (1+\epsilon)\alpha_{m,k}.$$

Theorem 1 proves that for large t, with high probability the degree distribution of $G_m^{(t)}$ follows a power law with exponent $\beta = 3$ (formally justifying the conclusions derived in [5]). The reader will note that Theorem 1 is stated as a concentration result for degrees in the range $0 \leq k \leq t^{1/15}$; as remarked in [14], this may be extended to degrees $k > t^{1/15}$. The power law exponent $\beta = 3$ is independent of the choice of m.

Bollobás, Riordan [12] prove the following theorem which computes the diameter of G_m^t.

Theorem 2. *Fix an integer $m \geq 2$ and a positive real number ϵ. With probability 1 as $t \to \infty$, G_m^t is connected and*

$$(1-\epsilon)\frac{\log t}{\log \log t} \leq diam(G_m^t) \leq (1+\epsilon)\frac{\log t}{\log \log t}.$$

A set of preferential attachment models different than the LCD model were proposed by Aiello et al. [3]. In [3], four evolving graph models were given. Three models produce directed graphs, while one generates undirected graphs. These models have some advantage over the LCD model, since the power law exponent β may roam over the interval $(2, \infty)$, dependent on the choice of parameters. Not surprisingly, these models are more complex in their description than the LCD model. We summarize only one such model that produces directed graphs, named Model C in [3].

The parameters are $m^{e,e}, m^{e,n}, m^{n,e}, m^{n,n} \in \mathbb{N}^+$ and a fixed finite directed graph H. (We adopt a simpler version of the parameter set in our description; in [3], the numbers are chosen according to some bounded probability distribution.) At time $t + 1$, add $m^{e,e}$ directed edges randomly among all nodes. The origins are chosen using preferential attachment with respect to the current out-degree and the destinations are chosen using preferential attachment with respect to the current in-degree. Add $m^{e,n}$ directed edges into v_{t+1} randomly. The origins are chosen using preferential attachment with respect to the current out-degree. Add $m^{n,e}$ directed edges from v_{t+1} randomly. The destinations are chosen using preferential attachment with respect to the current in-degree. Add $m^{n,n}$ directed loops to v_{t+1}.

The following result was proved in [3] using the Azuma-Hoeffding inequality (see Theorem 3 of [3] for a precise statement of the concentration results). Note that Model C produces directed graphs whose in- *and* out-degree distribution follow power laws.

Theorem 3. *For graphs generated by model C, with probability 1 as t tends to* ∞*, the out-degree distribution follows a power law with the exponent*

$$\beta = 2 + \frac{m^{n,n} + m^{n,e}}{m^{e,n} + m^{e,e}}.$$

With probability 1 as t tends to ∞*, the in-degree sequence follows a power law with exponent*

$$\beta = 2 + \frac{m^{n,n} + m^{e,n}}{m^{n,e} + m^{e,e}}.$$

There are other important preferential attachment models such as the model of Cooper and Frieze [28]. Their model is fairly complex, owing to its large number of parameters. Their proof of a power law degree distribution for graphs generated by their model is novel, since it uses martingale techniques along with the Laplace method for the solution of linear difference equations. Both Dorogovtsev et al. [32] and Drinea et al. [33] introduced a variation into preferential attachment where each node is assigned a constant *initial attractiveness am.* The probability that a new node is joined to an existing one u is proportional to its in-degree plus am. Buckley, Osthus [18] gave a rigorous version of this model along the lines of the LCD model. A model using preferential attachment to generate directed graphs in a way different than the ACL and LCD models was given in Bollobás et al. [13].

5 Off-line Web Graph Models

We discuss an interesting off-line model for the web introduced by Chung, Lu [24, 25]. The ER model $G(n, p)$ may be generalized as follows. Let $\mathbf{w} = (w_1, \ldots, w_n)$ be a *graphic* sequence; that is, the degree sequence of some graph of order n. We define a model for random graphs with expected degree sequence $\mathbf{w}$, written $G(\mathbf{w})$, as follows. The edge between v_i and v_j is chosen independently with probability p_{ij} where p_{ij} is proportional to the product $w_i w_j$. Then $G(n, p)$ may be viewed as a special case of $G(\mathbf{w})$ by taking $\mathbf{w}$ to equal the n-sequence $(pn, pn, \ldots, pn)$. In this way, Chung, Lu [24, 25] consider $G(\mathbf{w})$ where the expected degree sequence is a power law with fixed exponent β in the interval $(2, \infty)$. They refer to such $G \in G(\mathbf{w})$ as *power law random graphs*. The reader will note that the model $G(\mathbf{w})$ generates off-line graphs, unlike all the other models in this survey. The motivation for the study of power law random graphs comes in part from the fact that off-line models are easier to work with mathematically than on-line models. For instance, in contrast to off-line models, for on-line models the probability space for the random graph generated at time-step $t + 1$ is different than the one at time-step t.

In [24], the order of connected components of the graphs in $G(\mathbf{w})$ is investigated. The paper [25] proves the following result, which exposes a nice connection between a power law degree distribution and the small world property.

Theorem 4. *Suppose that* $G \in G(\mathbf{w})$ *has* n *nodes and expected degree sequence* $\mathbf{w}$ *following a power law with exponent* $\beta > 2$. *Let* G *have average degree* $d > 1$ *and maximum degree* m *satisfying*

$$\log m \gg \frac{\log n}{\log \log n}.$$

For all values of $\beta > 2$, *with probability* 1 *as* n *tends to* ∞, *the graph* G *is connected with*

$$diam(G) = \Theta(\log n).$$

If $2 < \beta < 3$, *then with probability* 1 *as* n *tends to* ∞,

$$L(G) \le (2 + o(1)) \left(\frac{\log \log n}{\log(1/(\beta - 2))} \right).$$

If $\beta = 3$, *then with probability* 1 *as* n *tends to* ∞,

$$L(G) = \Theta \left(\frac{\log n}{\log \log n} \right).$$

If $\beta > 3$, *then with probability* 1 *as* n *tends to* ∞,

$$L(G) = (1 + o(1)) \frac{\log n}{\log d}.$$

Expected power law degree sequences fall into the more general category of *admissible expected degree sequences* introduced in [25]. The results of Theorem 4 generalize to $G(\mathbf{w})$ with admissible expected degree sequences; see Theorems 1 and 2 of [25].

A recent paper of Chung, Lu [26] uses power law random graphs in the design of a certain off-line model named the *hybrid power law model*. This model generates so-called *hybrid graphs*, whose edge set is the disjoint union of a *global graph* and a *local graph*. The results of [26] show that hybrid graphs satisfy properties 2 and 3 of Section 3, and in addition, are locally highly connected.

6   Copying Models

We saw in Section 4 the connection between preferential attachment and power law degree distributions. In this section, we consider an evolving graph model that uses a paradigm different than preferential attachment, but nevertheless with high probability generates graphs with power law degree distributions. The *linear growth copying model* was introduced in Kleinberg et al. [41] and rigorously analyzed in Kumar et al. [43]. It has parameters $p \in (0,1)$, $d \in \mathbb{N}^+$, and a fixed finite directed graph H with constant out-degree d. Assume that G_t has constant out-degree d. At time $t + 1$, an existing node, which we refer to as u_t, is chosen u.a.r. from the set of all existing nodes. The node u_t is called the *copying node*. For each of the d out-neighbours w of u_t with probability p, add a directed edge (v_{t+1}, z), where z is chosen u.a.r. from $V(G_t)$, and with the remaining probability $1 - p$ add the directed edge (v_{t+1}, w). The authors of [43] use martingales and the Azuma-Hoeffding inequality to prove the following (see Theorems 8 and 9 of [43] for a precise statement of the concentration results).

Theorem 5. *With probability* 1 *as* t *tends to* ∞, *the copying model generates directed graphs* G_t *whose in-degree distribution converges to a power law with exponent*

$$\beta = \frac{2-p}{1-p}.$$

Property 4 of Section 3 (that is, the presence of many dense bipartite subgraphs) is a desirable property for graphs generated by a web graph model. Kumar et al. [43] analyze the model of Aiello et al. [2] (which was defined historically before the ACL models) and demonstrate that this model generates graphs which on average contain few bipartite cliques. Two subgraphs of a graph are *distinct* if they have distinct vertex sets. Let $K_{t,i,d}$ denote the expected number of distinct $K_{i,j}$'s which are subgraphs of G_t.

Theorem 6. *In the linear growth copying model with constant outdegree* d, *for* $i \leq \log t$,

$$K_{t,i,d} = \Omega(t \exp(-i)).$$

A new copying model $G(p, \rho, H)$ was recently introduced in Bonato, Janssen [16], motivated by the copying model, the generalized copying graphs of Adler, Mitzenmacher [1], and partial duplication model for biological networks in Chung et al. [23]. The three parameters of the model $G(p, \rho, H)$ are $p \in (0, 1)$, a monotone increasing *random link function* $\rho : \mathbb{N} \to \mathbb{N}$, and a fixed finite initial graph H. The new node v_{t+1} acquires its neighbours as follows. Choose an existing node u from G_t u.a.r.. For each neighbour w of u, independently add an edge from v_{t+1} to w with probability p. In addition, choose $\rho(t)$-many nodes from $V(G_t)$ u.a.r., and add edges from v_{t+1} to each of these nodes.

The existing research on models of W deals almost exclusively with finite graphs. However, in the natural sciences, models are often studied by taking the infinite limit. Limiting behaviour can clarify the similarities and differences between models, and show the consequences of the choices made in the model.

Limit behaviour of a deterministic copying model was investigated in [15], and limit behaviour of the $G(p, \rho, H)$ model was studied in [16]. For a positive integer n, a graph is *n-existentially closed* or *n-e.c.* if for each pair of disjoint subsets X and Y of nodes of G with $|X \cup Y| = n$, there exists a node $z \notin X \cup Y$ joined to every node of X and to no node of Y. A graph is *e.c.* if it is *n*-e.c. for all positive integers n. By a back-and-forth argument, any two countable e.c. graphs are isomorphic. The unique isomorphism type of countable e.c. graphs is the *infinite random graph* R. The graph R takes its name from the fact that for any fixed $p \in (0, 1)$, with probability 1, a graph $G \in G(\mathbb{N}, p)$ is e.c. The graph R has a rich structure, which the interested reader may read more about in the surveys of P. Cameron [20, 21].

If $(G_t : t \in \mathbb{N})$ is a sequence of graphs with G_t an induced subgraph of G_{t+1}, then define the *limit* of the G_t, written

$$G = \lim_{t \to \infty} G_t,$$

by

$$V(G) = \bigcup_{t \in \mathbb{N}} V(G_t), \ E(G) = \bigcup_{t \in \mathbb{N}} E(G_t).$$

The following result is essentially stated in [16].

Theorem 7. *Fix $p \in (0, 1)$, H, and $\rho = \lfloor \alpha t^s \rfloor$, where α and s are non-negative real numbers with $\alpha, s \in [0, 1]$ and $\alpha + p < 1$. Let $G = \lim_{t \to \infty} G_t$ be generated according to the model $G(p, \rho, H)$.*

1. *If $s = 1$ and $\lfloor \alpha t^s \rfloor \geq 1$ for all $t > 0$, then with probability 1 G is isomorphic to R.*
2. *If $s \in [0, 1)$ and $\lfloor \alpha t^s \rfloor \geq 1$ for all $t > 0$, then with probability 1 G is $\lfloor \frac{1}{1-s} \rfloor$-e.c.*
3. *If $s \in [0, 1)$, then with positive probability G is not isomorphic to R.*

Theorem 7 presents an example of threshold behaviour for convergence to R: with high probability, as s tends to 1, the limit G acquires more and more properties of R, but with positive probability is not itself isomorphic to R. At $s = 1$, we obtain R with high probability.

A new deterministic model for the web graph of note is the *Heuristically Optimized Trade-offs* or *HOT* model of Fabrikant et al. [36]. In the HOT model, nodes correspond to points in Euclidean space, and each node u will link to the node v that performs best in terms of an optimization function which is a linear combination of proximity between u and v, and centrality of v in the network. This implicitly suggests some degree of copying behaviour: a new node whose position is very close to that of an existing node, will have a similar optimization function and hence, is likely to connect to the same node.

7   Growth-Deletion Models

In all of the models we presented in Sections 4 and 6, at each time step nodes and edges are added, but never deleted. An evolving graph model incorporating in its design both the addition and deletion of nodes and edges may more accurately model the evolution of the web graph. One approach to this was adopted by Bollobás, Riordan [11], who consider the effect of deleting a set of nodes *after* nodes have been generated in the LCD model. The purpose of this study was to investigate the robustness of graphs generated by the LCD model to random failures, and the vulnerability of these graphs to random attack. We now describe two recent models, developed independently of each other, that incorporate the addition and deletion of nodes *during* the generation of nodes. We refer to such models as *growth-deletion models*.

We first describe the growth-deletion model of Chung, Lu [27]. They introduce a model $G(p_1, p_2, p_3, p_4, m)$, with parameters m a positive integer, and probabilities p_1, p_2, p_3, p_4 satisfying $p_1 + p_2 + p_3 + p_4 = 1$, $p_3 < p_1$, and $p_4 < p_2$; the graph H is a fixed nonempty graph. To form G_{t+1}, we proceed as follows. With probability p_1, add v_{t+1} and m edges from v_{t+1} to existing nodes chosen by preferential attachment. With probability p_2, add m new edges with endpoints to be chosen among existing nodes by preferential attachment. With probability p_3, delete a node chosen u.a.r. With probability p_4, delete m edges chosen u.a.r.

By coupling with off-line random graphs, Chung, Lu [27] prove the following result.

Theorem 8.   *1. With probability 1 as $t \to \infty$, the degree distribution of a graph G_t generated by $G(p_1, p_2, p_3, p_4, m)$ follows a power law distribution with exponent*

$$\beta = 2 + \frac{p_1 + p_2}{p_1 + 2p_2 - p_3 - 2p_4}.$$

2. Suppose that $m > \log^{2+\epsilon} n$. For $p_2 < p_3 + 2p_4$, we have $2 < \beta < 3$. With probability 1 as $t \to \infty$, G_t is connected with

$$diam(G_t) = \Theta(\log t)$$

and

$$L(G_t) = O\left(\frac{\log \log t}{\log(1/(\beta - 2))}\right).$$

3. Suppose that $m > \log^{2+\epsilon} n$. For $p_2 \geq p_3 + 2p_4$, we have $\beta > 3$. With probability 1 as $t \to \infty$, G_t is connected with

$$diam(G_t) = \Theta(\log t)$$

and

$$L(G_t) = O\left(\frac{\log t}{\log d}\right),$$

where d is the average degree of G_t.

Another recent growth-deletion model developed independently of [27] is the one of Cooper et al. [29]. The parameters for this model are fixed p_1 and p_2 in $(0, 1)$ satisfying $p_2 \leq p_1$, and H is K_1. With probability $1 - p_1$ delete a node of

G_{t-1} chosen u.a.r. If G_{t-1} has no nodes, then do nothing. With probability p_2, add m edges from v_{t+1} joined to existing nodes chosen by preferential attachment. The graph is made simple by deleting multiple edges. If there are no edges nor nodes in G_{t-1}, then begin again at time $t = 0$. If there are no edges but some nodes in G_{t-1}, then add v_{t+1} joined to an existing node chosen u.a.r. With probability $p_1 - p_2$, add m edges between existing nodes, with endpoints chosen by preferential attachment. The graph is made simple by deleting multiple edges and deleting any loops. If there are no edges nor nodes in G_{t-1}, then begin again at time $t = 0$. If there are no edges but some nodes in G_{t-1}, then do nothing.

Let $D_k(t)$ be the number of nodes of degree $k \geq 0$ in G_t, and let $\mathbb{E}(D_k(t))$ be the expectation of this random variable. Let

$$\gamma = \frac{2p_1}{3p_1 - 1 - p_2} \text{ and } \rho = \frac{p_2}{p_1}.$$

Cooper et al. [29] prove the following.

Theorem 9. *Assume that $p_1 + p_2 > 1$. Then there exists a constant $C = C(m, p_1, p_2)$ such that for $k \geq 1$ and $1/2 < p_1 \leq 1$,*

$$\left| \frac{\mathbb{E}(D_k(t))}{t} - Ck^{-1-\gamma} \right| = O\left(t^{-\rho/8}\right) + O(k^{-2-\gamma}).$$

As noted in [29], with a suitable choice of p_1 and p_2, γ may take any value in the interval $(1, \infty)$, and so there is a power law for this model with exponent $\beta = 1 + \gamma \in (2, \infty)$.

An intriguing problem is to rigorously analyze the degree distributions of growth deletion models where the choice of nodes and edges to delete is not made u.a.r. A recent model of Flaxman et al. [37] considers an *adversarial* growth deletion model, and analyzes the size of the connected components of graphs generated by the model.

References

1. A. Adler, M. Mitzenmacher, Towards compressing web graphs, In: *Proceedings of the Data Compression Conference*, 2001.
2. W. Aiello, F. Chung, L. Lu, A random graph model for massive graphs, *Experimental Mathematics* **10** (2001) 53-66.
3. W. Aiello, F. Chung, L. Lu, Random evolution in massive graphs, *Handbook on Massive Data Sets* (James Abello et al. eds.), Kluwer Academic Publishers, (2002), 97-122.
4. R. Albert, A. Barabási, H. Jeong, Diameter of the World-Wide Web, *Nature* **401** (1999) 130.
5. R. Albert, A. Barabási, Emergence of scaling in random networks, *Science* **286** (1999) 509-512.
6. P. Baldi, P. Frasconi, P. Smyth, *Modeling the Internet and the Web, Probabilistic Methods and Algorithms*, John Wiley & Sons, Ltd, Chichester, West Sussex, England, 2003.

7. A. Barabási, *Linked: How Everything Is Connected to Everything Else and What It Means*, Perseus Publishing, Cambridge MA, 2002.
8. B. Bollobás, *Modern Graph Theory*, Springer-Verlag, New York 1998.
9. B. Bollobás, *Random graphs, Second edition,* Cambridge Studies in Advanced Mathematics **73** Cambridge University Press, Cambridge, 2001.
10. B. Bollobás, O. Riordan, Mathematical results on scale-free graphs, *Handbook of graphs and networks* (S. Bornholdt, H. Schuster eds.), Wiley-VCH, Berlin (2002).
11. B. Bollobás, O. Riordan, Robustness and vulnerability of scale-free random graphs, *Internet Mathematics* **1** (2004) 1-35.
12. B. Bollobás, O. Riordan, The diameter of a scale-free random graph, *Combinatorica* **24** (2004) 5-34.
13. B. Bollobás, C. Borgs, T. Chayes, O. Riordan, Directed scale-free graphs, submitted.
14. B. Bollobás, O. Riordan, J. Spencer, G. Tusnády, The degree sequence of a scale-free random graph process, *Random Structures Algorithms* **18** (2001) 279-290.
15. A. Bonato, J. Janssen, Infinite limits of copying models of the web graph, *Internet Mathematics* **1** (2004) 193-213.
16. A. Bonato, J. Janssen, Limits and power laws of models for the web graph and other networked information spaces, accepted to the *Proceedings of Combinatorial and Algorithmic Aspects of Networking*, 2004.
17. A. Broder, R. Kumar, F. Maghoul, P. Raghavan, S. Rajagopalan, R. Stata, A. Tomkins, J. Wiener, Graph structure in the web, *Computer Networks* **33** (2000) 309-320.
18. P.G. Buckley, D. Osthus, Popularity based random graph models leading to a scale-free degree sequence, submitted.
19. G. Caldarelli, P. De Los Rios, L. Laura, S. Leonardi, S. Millozzi, A study of stochastic models for the Web Graph, Technical Report 04-03, dipartimento di Informatica e Sistemistica, Universita' di Roma "La Sapienza", 2003.
20. P.J. Cameron, The random graph, *Algorithms and Combinatorics* (R.L. Graham and J. Nešetřil, eds.), Springer Verlag, New York (1997) 333-351.
21. P.J. Cameron, *The random graph revisited*, in: European Congress of Mathematics, Vol. I (Barcelona, 2000), 267–274, Progr. Math., 201, Birkhäuser, Basel, 2001.
22. S. Chakrabarti, *Mining the Web, Discovering Knowledge from Hypertext Data*, Morgan Kauffman Publishers, San Francisco, 2003.
23. F. Chung, G. Dewey, D.J. Galas, L. Lu, Duplication models for biological networks, *Journal of Computational Biology* **10** (2003) 677-688.
24. F. Chung, L. Lu, Connected components in random graphs with given degree sequences, *Annals of Combinatorics* **6** (2002) 125-145.
25. F. Chung, L. Lu, The average distances in random graphs with given expected degrees, *Internet Mathematics* **1** (2004) 91-114.
26. F. Chung, L. Lu, The small world phenomenon in hybrid power law graphs, *Complex Networks*, (Eds. E. Ben-Naim et. al.), Springer-Verlag (2004) 91-106.
27. F. Chung, L. Lu, Coupling on-line and on-line analyses for random power law graphs, submitted.
28. C. Cooper, A. Frieze, On a general model of web graphs, *Random Structures Algorithms* **22** (2003) 311–335.
29. C. Cooper, A. Frieze, J. Vera, Random deletions in a scale free random graph process, submitted.
30. R. Diestel, *Graph theory*, Springer-Verlag, New York, 2000.
31. S.N. Dorogovtsev, J.F.F. Mendes, *Evolution of networks: ¿From biological nets to the Internet and WWW*, Oxford University Press, Oxford, 2003.

32. S.N. Dorogovtsev, J.F.F. Mendes, A.N. Samukhin, Structure of growing networks with preferential linking, Physical Review Letters **85** (2000) 4633-4636.
33. E. Drinea, M. Enachescu, M. Mitzenmacher, Variations on random graph models for the web, technical report, Department of Computer Science, Harvard University, 2001.
34. P. Erdős, A. Rényi, On random graphs I, *Publ. Math. Debrecen* **6** (1959) 290–297.
35. P. Erdös, A. Rényi, On the evolution of random graphs, *Publ. Math. Inst. Hungar. Acad. Sci.* **5** (1960) 17–61.
36. A. Fabrikant, E. Koutsoupias, C. Papadimitriou, Heuristically optimized Trade-offs: a new paradigm for power laws in the internet, In: *Proceedings of the 34th Symposium on Theory of Computing*, 2002.
37. A. Flaxman, A. Frieze, J. Vera, Adversarial deletions in a scale free random graph process, submitted.
38. E.N. Gilbert, Random graphs, *Annals of Mathematical Statistics* **30** (1959) 1141-1144.
39. M.R. Henzinger, Algorithmic challenges in web search engines, *Internet Mathematics* **1** (2004) 115-126.
40. S. Janson, T. Luczak, A. Ruciński, *Random Graphs*, John Wiley and Sons, New York, 2000.
41. J. Kleinberg, S.R. Kumar, P. Raghavan, S. Rajagopalan, A. Tomkins, The web as a graph: Measurements, models and methods, In: *Proceedings of the International Conference on Combinatorics and Computing*, **1627** in LNCS, Springer-Verlag, 1999.
42. R. Kumar, P. Raghavan, S. Rajagopalan, A. Tomkins, Trawling the web for emerging cyber-communities, In: *Proceedings of the 8th WWW Conference*, 1999.
43. R. Kumar, P. Raghavan, S. Rajagopalan, D. Sivakumar, A. Tomkins, E. Upfal, Stochastic models for the web graph, In: *Proceedings of the 41th IEEE Symp. on Foundations of Computer Science*, 57-65, 2000.
44. R. Kumar, P. Raghavan, S. Rajagopalan, D. Sivakumar, A. Tomkins, E. Upfal, The web as a graph, In: *Proc. 19th ACM SIGACT-SIGMOD-AIGART Symp. Principles of Database Systems*, Publ., Dordrecht, 2002.
45. M. Mitzenmacher, A brief history of generative models for power law and lognormal distributions, *Internet Mathematics* **1** (2004) 226-251.
46. S.H. Strogatz, D.J. Watts, Collective dynamics of 'small-world' networks, *Nature* **393** (1998) 440-442.

You Can Get There from Here: Routing in the Internet

Jean-Charles Grégoire[1] and Angèle M. Hamel[2]

[1] Institut International des Télécommunications,
800 de la Gauchetière Ouest, Bureau 6700,
Montréal, Québec, H5A 1K6, Canada
and Institut National de la Recherche Scientifique–Énergie,
Matériaux, et Télécommunications,
800 de la Gauchetière Ouest, Bureau 6900,
Montréal, Québec, H5A 1K6, Canada
`Jean-Charles.Gregoire@iitelecom.com`
[2] Department of Physics and Computer Science,
Wilfrid Laurier University, Waterloo, Ontario, N2L 3C5, Canada
`ahamel@wlu.ca`

Abstract. Routing in the Internet is performed with the help of IP forwarding lookup tables; however, searching these tables efficiently for the "next hop" is nontrivial, and complex data structures make an important contribution in this regard. This article surveys the three main approaches to the problem and investigates the associated data structures and algorithms. We also explore variations on these approaches and conclude with a discussion of current trends.

1 Introduction

The main design feature of the Internet is reliability. Through a combination of mesh topology and distributed computation of end-to-end paths, there is no single point of failure. Routers, the Internet's communication relays, have multiple connectivity and exchange through routing protocols so that they can compute the "best" routes to destination. The information they exchange supports their forwarding operations, and they act like switches, taking packets from one interface and forwarding them on to another. The decision of the right interface for forwarding is made with the help of lookup tables, and it is important that this decision be performed extremely fast. However, a simple linear search, or even binary or hash–based search, of these tables is not feasible due to their size and the speed with which the router must make a decision. Therefore, complex data structures have been invented to optimize this process. This survey examines a number of these data structures.

The forwarding done by the router is based on longest prefix matching. IP addresses (in IPv4) are 32 bits long and consist of a variable length, mask defined, network portion called the *prefix*, and a host portion. The network portion

A. López-Ortiz and A. Hamel (Eds.): CAAN 2004, LNCS 3405, pp. 173–182, 2005.
© Springer-Verlag Berlin Heidelberg 2005

specifies the network that contains the host, while the host portion specifies a particular host on that network. The strategy employed by the router is to search a forwarding table for the longest prefix that matches the address the packet wants to reach and then send the packet to the corresponding "next hop" address. In practice, addresses are aggregated, or combined, to reduce the size of the table. But even with this improvement the table space can be too large to search efficiently. A further complication is the updating of the table—the Internet is in constant flux as some routers and hosts join and others drop out. The time to refresh or update the forwarding table is an additional concern. To summarize, the lookup operation must be fast, and the underlying data structure must be small and easy to update or refresh. We must also keep in mind that this operation is embedded in hardware in most modern, performance–critical routers and caching is not an option because of lack of locality. On the other hand, the table is typically computed by another processor based on updates generated by routing algorithms.

The classical approach to the problem of router lookup is a variation on the Patricia trie. An ordinary trie is a m–ary tree data structure with nodes that contain strings of characters [8]. For prefix matching we use a binary trie in which each node at level i is a string of length $i - 1$ ones and zeros, and each node has two children—one child is labeled by the parent string concatenated with one, and the other child is labeled by the parent string concatenated with zero. Certain of these nodes will be identified as prefixes on the router. A search for a string through this tree will follow a path from root to leaf, where the leaf represents the string. The last prefix encountered on this path will be the longest matching prefix. The trie–based approach is actually optimal, as can be shown using finite state automata associated to the routing table [4].

A Patricia (Practical Algorithm To Retrieve Information Coded in Alphanumeric) trie is a first stage improvement on an ordinary trie—it introduces rudimentary compression by collapsing nodes with a single child and recording the

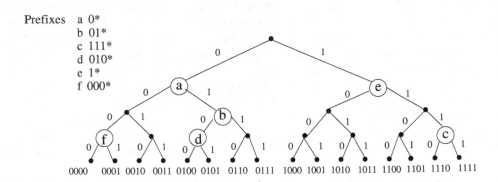

Fig. 1. Given a four-bit pattern, find the longest matching prefix by searching the tree, starting at the root and finishing at the appropriate leaf. The search describes a root–to–leaf path. The lowest prefix on this path is the longest matching prefix. The figure is patterned after an example in Ruiz–Sanchez *et al.* [16]

number of such nodes "skipped." However, Patricia tries are not suited to longest prefix matching because they may inadvertently skip the prefix required. For example, suppose we are searching for a string with prefix 1001111 and suppose the Patricia trie is constructed such that it skips the three 1's in positions 4,5, and 6 and leads directly to a prefix of 1001110. Then this is not a match, so it backtracks. For certain choices, the backtracking overhead can be excessive [19].

Excellent surveys of these topics can be found in [16] and [19].

2   Data Structures for Forwarding

Optimizing the forwarding process emerged as a critical problem in the nineties with the explosive growth of both the size of the Internet and the capacity of communication links. Breakthroughs in this area are thus fairly recent. We present in this section three main approaches, respectively based on tries, ranges, and hashing.

2.1   Three Approaches

Three papers are milestones in terms of improved data structures: Degermark et al. [2], Lampson et al. [9], and Waldvogel et al. [20].

The structure of Degermark [2] is a multibit trie (also called a Lulea trie) that provides a high degree of compression, substantially reducing the size of the data structure while also minimizing the number of memory accesses. The downside to this approach, however, is that updating can be a costly procedure, requiring modifications of substantial portions of the trie.

The Degermark data structure can be pictured initially as a binary tree with 2^{32} leaves—one for each possible IP address in IPv4. However, this is then compressed into a fixed–stride trie with strides of sizes 16, 8, and 8. A stride of size k means a binary tree of depth k is replaced by a single node with 2^k children—one for each leaf in the original binary tree. Suppose we have an address of length 2^{32} whose first k bits match exactly a prefix on the list. Then this prefix will be one of the 2^k children, and this child can contain a listing of this prefix. Suppose we have an address whose first k bits match the first k bits of a prefix but the prefix itself is longer than k bits. Then the first k bits will match one of the 2^k children and the child will contain a pointer to the next set of bits.

Degermark et al. further refine this broadbased scheme by "leaf pushing" all the prefix information to the leaves while retaining the pointer information at the nodes. Thus internal nodes contain only pointer information and leaves contain only prefix information. Further, if a prefix is associated with consecutive positions in the node, then bitmap compression is used to replace the repeated positions with zeros. The computation is further speeded up with the introduction of other arrays.

The Lampson approach [9] is a prefix range approach and takes advantage of the fact that a prefix (usually) represents a block or range of IP addresses. Instead of searching for an individual address, we can search for the range that

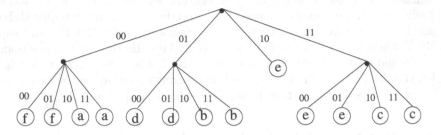

Fig. 2. Fixed stride trie with strides of size two and the prefixes pushed to the leaves. The figure is patterned after an example in Ruiz–Sanchez *et al.* [16]

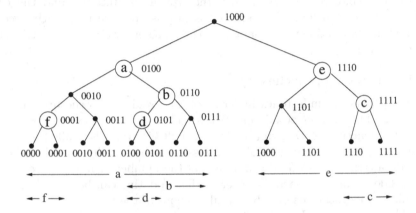

Fig. 3. A binary range tree. The nodes are arranged so that the leaf siblings correspond to endpoints of range intervals. The figure is patterned after an example in Ruiz–Sanchez [16]

contains that address, or, more properly, search for the endpoints of the range. Thus the address space is partitioned into ranges, and a binary search is done on the ranges. If a value falls into more than one range, as would be the case if one range were contained wholly within another, then it is matched to the shorter prefix range. For each range endpoint we keep track of two pieces of next hop information—the next hop if the element is strictly greater than the endpoint and the next hop if the element is equal to the endpoint. Geometrically we can picture this as a tree with IP addresses at the leaves. This approach can be substantially improved through the use of a multiway range tree, which is a tree in which each node has associated to it a range called an address span which corresponds to the IP addresses that lie in the leaves below it in the tree. Then each prefix corresponds to a range, and we associate a prefix to a node if the address span for that node is contained within the prefix range (and the address span of the node's parent is not). Then we search the tree in the usual way. By placing ranges at nodes we have essentially compressed the tree [21].

The Waldvogel scheme [20] is a hash–based approach. The set of prefixes is partitioned into hash tables where there is one hash table for each possible prefix length. The approach is to do a binary search on the prefix lengths, then examine the hash table for that length to find the particular prefix. If the prefix is found, then the search continues among longer prefix lengths (as there might be a longer matching prefix). If the prefix is not found, then the search reverses to continue among shorter prefix lengths.

However, there is a potential problem with this approach, namely if a prefix does not match at a particular length yet there is a match for a longer length. The paper solves this problem by introducing a marker that sends the search towards longer prefixes if required. For example, adapting an example of Varghese [19], suppose our prefix table has three prefixes: $\{1*, 10*, 10011*\}$. Then we have five hash tables: one for lengths 1, 2, and 5—each containing one prefix—and two empty hash tables for lengths 3 and 4. Then a binary search for prefix 1001 starts at length 3 which is empty so, by the description in the previous paragraph, the search should move towards smaller length prefixes. Yet an exact match is in length 5. Then the introduction of a marker 100 at length 3 will force the search towards longer prefixes. A final difficulty remains, however, as the markers can bias the search towards longer prefixes so that if just part of the prefix matches, the marker will send the search towards longer prefixes when it might actually more properly be sent towards shorter prefixes. In our example this is what would happen if we were searching for 100–it actually matches 10* but the marker would send it towards 10011*. The solution to the problem is a pre–computed value which is the longest prefix that matches the marker. If a search further towards longer prefixes fails to find a match, the default is to use this pre–computed value, thus avoiding a search towards shorter prefixes as well.

Complexity. Ruiz–Sanchez *et al.* [16] contains a detailed comparison of the complexity of the various popular approaches. Let N be the number of prefixes in the router table and let W be the length of the longest prefix in the table. The simplest binary trie has a worst case lookup of $O(W)$ (as the height of the trie will be the length of the longest prefix), update cost of $O(W)$, and a memory requirement of $O(NW)$. The Degermark approach improves lookup to $O(W/k)$ in the worst case, where k is the maximum stride size in bits, and the memory requirement is $O(2^k NW/k)$. The update cost is uncalculated by Ruiz–Sanchez as leaf pushing makes updating difficult and impractical. The Waldvogel scheme comes with a $O(\log_2 W)$ worst case, with an update cost of $O(N \log_2 W)$ and memory requirement in $O(\log_2 W)$—these last two are strongly influenced by the fact that $\log_2 W$ markers may be needed for each prefix. The range search approach of Lampson is also logarithmic, but in N, $O(\log_2 N)$ in the worst case (due to binary search on intervals), with $O(N)$ for update cost and $O(N)$ for memory, which are the costs of recomputing intervals and storing endpoints.

2.2   Recent Improvements

The three approaches outlined in the previous section all appeared around the same time. Since then a number of improvements on these schemes have been suggested. Each of the three schemes suffers from a primary drawback. In the case of tries, updates are difficult as the entire trie may have to be modified. In the case of ranges, speed and memory both suffer and updating can also be problematic in some implementations. In the case of hashing, collisions need to be accounted for. Here we explore a sampling of some of the trends in solving these problems.

Tries. One of the problems with the Degermark approach is that leaf pushing makes the trie difficult to update, as the update may possibly flow the height of the tree. An improved scheme called tree bitmap and due to Eatherton *et al.* [5] allows for incremental updates within a fixed–stride trie. It begins with the same principle as Degermark, but instead of leaf pushing it uses two bitmaps per node to keep track of information. One bitmap keeps track of internally stored prefixes (the elements Degermark pushes to the leaves); the other bitmap keeps track of pointers to other nodes in the trie. This scheme also uses smaller strides (e.g. 4 bit instead of 8 or 16) and this results in only one memory access per node instead of two.

Another approach to improved trie updating is developed by Pao and Li [15] for the LC–trie [13]. The level compressed, or LC, trie is a structure similar to the Degermark trie except that the Degermark trie features additional bitmap compression. The LC–trie features level compression (the trading of levels/depth in the trie for increased numbers of children at the nodes) and leaf pushing. The Pao and Li approach consists in relocating some of the prefixes from the leaves to the internal nodes and allowing insertion to possibly increase the degree, or branch factor, at a node. These improvements allow thousands of updates per second as opposed to a few updates per second as allowed by the ordinary LC trie.

Kim and Choi [7] also identify that one of the most severe problems with tries is how updates affect several levels of the trie. They further identify that compression, as found in the Degermark scheme, while superior in terms of memory usage, actually leads to substantial problems in terms of updating as it can be difficult to change a compressed table. Their solution is to amend the nodes of a trie to contain pointers *and* next hop information instead of just one or the other. While this does increase the memory immediately at the node, it leads to fewer level modifications when updating and also to fewer memory accesses. Indeed, their goal is to reduce the update time so that an update could be completed in the time it takes between lookups. Their scheme also lends itself to implementation using hardware pipelining.

Sundström and Larzon [18] propose a trie-based data structure with a different framework. Although their structure begins as a variable–stride trie, the first few layers of nodes contain pointers to objects called block trees instead of pointers to other nodes or leaves. A block tree is actually a generalization of the multiway range structure in Lampson *et al.* [9]. It is usual in tries for nodes to contain pointers to subtrees so that a search through the trie progresses from a node through the pointer into the subtree. However, in a block tree, nodes have blocks associated to them instead of pointers, and these blocks contain the subtree information. This results in an implicit data structure in which the information is accessed by way of the manner in which is is stored, rather than by way of pointers. It also results in very few memory accesses. Significantly, the Sundström and Larzon scheme does not suffer from the update problem of Degermark. In fact, it has performance superior to the tree bitmap of Eatherington [5].

Ranges. Chan *et al.* [1] propose a variation on the range–based algorithm. Their starting point is routing intervals similar to Lampson [9]. However, these are difficult to update. To solve this, Chan proposes a scheme to precompute the prefix corresponding to each range. Lampson *et al.* also precompute; however, their approach updates poorly, while the scheme of Chan, which combines a sorting and stack processing step of Lampson, updates efficiently. Chan further improves the multiway search tree structure of Warkhede *et al.* [21] by increasing branches and decreasing height. The resulting search cost is $O(\log_{16}(2N))$ and the resulting memory is $O(N)$, where N is the number of prefixes.

Sahni and Kim [17] develop a complex and intricate data structure to address the problem of fast updates in range–based methods. They construct a collection of binary search trees (best implemented as red–black trees), one of which is a basic interval tree whose leaves correspond to intervals. The other BST are prefix trees—there is one tree for each prefix P, and each tree consists of a root and a set of nodes corresponding to the longest prefix whose range includes the range of P. We search the basic interval tree first and in some instances the longest matching prefix is found there and in some instances we are pointed on to one of the prefix trees. The strength of this scheme is that the prefix matching as well as insertion and deletion of prefixes can be done in $O(\log N)$ time, where N is the number of prefixes.

Hashing. Hashing is obviously an attractive solution to the matching problem because it produces small, fixed–size outputs from longer, variable–sized inputs. However, in the case of longest prefix matching it is unclear just what the input should be: how much of the original address corresponds to the prefix we are looking for, especially since we want the longest such prefix? As already discussed, the Waldvogel scheme employs hashing. The paper of Lim *et al.* [11] also employs hashing but provides a scheme that additionally deals with collisions by introducing subtables as well as main hash tables. The main hash tables contain a hash and a corresponding initial set of prefixes. If a hash points to an entry in the main table, the prefix corresponding to that entry is compared to the prefix in the original address. If they do not match, a collision has occurred and a corresponding subtable is (binary) searched.

Another hash–based improvement on Waldvogel [20] can be found in Ioannidis *et al.* [6]. Their approach is to recognize that there may be extra memory available in the router and that additional hash tables can be created using this memory. Which additional hash tables should be built? This depends very much on the specific lookup table, but their scheme concentrates on areas that are frequently accessed. They formulate the problem as a knapsack problem and then derive a greedy approximation algorithm for it.

Different Approaches. In addition to improvements to the three primary approaches to the prefix matching problem, a number of different ideas have been proposed.

A radically different approach can be found in Dharmapurikar *et al.* [3] which uses Bloom filters to perform longest prefix matching. A Bloom filter is a vector–based data structure that efficiently stores a set of messages by using a set of hash functions to "program" the filter with a series of 1's corresponding to the hash function values. We begin with a zero vector of length n. Take k hash functions whose output is between 1 and n, and hash the message using each hash function. Then take the k outputs and change the corresponding bits in the vector to 1. Repeat this for each message. If the hash function output calls for a bit to be changed to a 1 and it is already a 1, then there is no change and it remains a 1. The final vector is the Bloom filter. Then a similar process is used to query the filter to see if a particular message is contained within. Obviously we can only determine the presence of a certain message to some probability. The Dharmapurikar approach associates a Bloom filter to the set of prefixes of a given length. It also maintains a hash table for each length, consisting of the prefixes of that length and the next hop information. To search for a given IP address, we first search the Bloom filters in parallel to obtain a list of possible matches and then search the associated hash table to find a definite match if one exists.

The paper of Liang *et al.* [10] changes the longest prefix matching problem into an only prefix matching problem. This approach has been taken by other algorithms as well. For example, one way to do this is to eliminate prefix overlap by extending prefixes. For example, suppose we have two prefixes, 11011* and 110*. Then replace 110* by 11010*, 11001*, 11000*, and now we can search for the only prefix that matches among {11011∗, 11010∗, 11001∗, 11000∗}. The authors first construct a binary trie in such a way as to partition the prefixes into nonoverlapping sets. They then define a parallel lookup framework that examines these nonoverlapping sets in parallel.

A further unique approach is the geometric approach of Pelegrini and Fusco [14]. They translate the longest prefix match problem into the geometric problem of probing a line divided into segments to find the shortest segment to which a particular point belongs. A standard approach to this geometric problem is to partition the line into a grid of buckets and then systematically search the buckets for the point. Here the authors build a tree data structure to encode the geometric problem and perform a number of optimizing refinements on it to improve performance. Their emphasis is on demonstrating the practicality of a geometric approach and their experiments show performance comparable to existing techniques.

A final novel approach is due to Mahramian *et al.* [12] who employ neural networks using a back propogation algorithm. This represents an interesting first step, although the authors admit training time and updating are problems with this scheme as is typical for neural networks.

3   Current Trends

On a practical note, progress in communication speeds drives the needs for faster lookup. 10Gbps (OC192) interfaces are commercially available today, and

40Gbps (OC768) is around the corner. At such speeds, lookup has to be performed within a few nanoseconds to allow the smallest packets to be processed at wire speed.

Beyond performance issues, memory is another significant constraint. The Internet continues to grow, and in spite of constant improvements in capacity, processing speeds require the use of SRAM technology, which is more expensive that the typical DRAM we find in Personal Computers. This is why so much effort has been put into improving the compression of information achieved with tries, although today's performance of classical algorithms appears quite adequate [19].

Compression, however, is achieved at the cost of ease of incremental update. Whereas incremental update can be done without interruption of forwarding, complete update would lead to disruptive slow down. The only alternative then is to use two memory banks, current and updated, which obviously doubles the memory requirements!

Further, there is always the issue of proprietary vs. licensed, or public technology. It can often be in the interest of manufacturers to own the technology they use, and the need for originality will remain important.

Besides IPv4 addresses, we can also look at forwarding for ethernet and IPv6. Ethernet switches use MAC identifiers to forward frames between switches. In this case, there is no hierarchy in the identifier space, or no longest prefix. A fixed-length prefix identifies the manufacturer, but assignment of identifiers to chips, and hence interface cards, is done chronologically, and these pieces of equipment can be arbitrarily spread across the network. Some of the trie-based approaches presented here could apply.

IPv6's level of deployment remains confidential at this point, although manufacturers' degree of commitment to this new(er) network protocol is getting stronger. Already most major computer platforms support it to a large degree, but progress is slower on the router side, and part of the reason is obvious. It is clear that hardware optimized for 32 bit addresses cannot automatically be used for 128 bit addresses, and further, incremental techniques based on strides will perform more slowly because they would require more steps. Obviously addresses are larger but there is a silver lining here as they are more systematically deployed geographically, unlike a large chunk of the IPv4 address space, and we can suppose that there will be a larger degree of prefix sharing. But since most of the academic research has been validated based on copies of routing tables from the core of the Internet, we will need a significant level of deployment before we have access to reliable data to validate such assumptions.

References

1. C.-T. Chan, P.-C. Wang, S.-C. Hu, C.-L. Lee, R.-C. Chen, High performance IP forwarding with efficient routing–table update, *Computer Comm.* 26 (2003), 1681–1692.
2. M. Degermark, A. Brodnick, S. Carlesson, and S. Pink, Small forwarding tables for fast routing lookups, *Proceedings of ACM SIGCOMM*, 1997, 3–14.

3. S. Dharmapurikar, P. Krishnamurthy, D.E. Taylor, Longest prefix matching using Bloom filters, *Proceedings of SIGCOMM*, 2003.
4. A. Donnelly and T. Deegan, IP router lookups as string matching, *Proceedings of IEEE Local Computer Networks (LCN)*, 2000.
5. W. Eatherton, G. Varghese, Z. Dittia, Tree bitmap: Hardware/sofware IP lookups with incremental updates, *Computer Comm. Review* 34 (2004), 97–122.
6. I. Ioannidis, A. Grama, M. Atallah, Adaptive data structures for IP lookups, *Proceedings of IEEE INFOCOM*, 2003.
7. B.-Y. Kim, Y.-H. Choi, A pipelined routing lookup scheme with fast updates, *Computer Comm.* 26 (2003), 1594–1601.
8. D.E. Knuth, *The Art of Computer Programming*, Vol. III, Addison–Wesley, 1973.
9. B.Lampson, V. Srinivasan, G. Varghese, IP lookups using multiway and multicolumn search. In *Proceedings of IEEE INFOCOM*, 1998.
10. Z. Liang, K. Xu, J. Wu, A scalable parallel lookup framework avoiding longest prefix match, ICOIN 2004, LNCS 3090, 2004 , 616–625.
11. H. Lim, J.-H. Seo, Y.-J. Jung, High speed IP address lookup architecture using hashing, *IEEE Comm. Letters* 7 (2003), 502–504.
12. M. Mahramian, N. Yazdani, K. Faez, H. Taheri, Neural network based algorithms for IP lookup and packet classification, *EurAsia-ICT 2002*, LNCS 2510, 204–211.
13. S. Nilsson and G. Karlsson, Fast address lookup for Internet routers, *Proceedings of IEEE Broadband Communications*, 1998.
14. M. Pellegrini and Giordano Fusco, Efficient IP table lookup via adaptive stratified trees with selective reconstructions, *Proceedings of ESA*, 2004, LNCS 3221, 772–783.
15. D. Pao and Y.-K. Li, Enabling incremental updates to LC–trie for efficient management of IP forwarding tables, *IEEE Comm. Letters* 7 (2003), 245-247.
16. M.A. Ruiz–Sanchez, E.W. Biersack, W. Dabbous, Survey and taxonomy of IP address lookup algorithms, *IEEE Network*, March/April 2001, 8–23.
17. S. Sahni and K.S. Kim, An $O(log\ n)$ dynamic router–table design, *IEEE Trans. on Computers*, 53 (2004), 351–363.
18. M. Sundström and L.-A. Larzon, High performance longest prefix matching supporting high–speed incremental updates and guaranteed compression, in *Proceedings of INFOCOM*, 2005.
19. G. Varghese, *Network Algorithmics*, Amsterdam: Elsevier, 2005.
20. M. Waldvogel, G. Varghese, J. Turner, and B. Plattner, Scalable high–speed IP routing lookups. *ACM Transactions on Computer Systems*, 19 (2001), 440–482.
21. P. Warkhede, S. Suri, G. Varghese, Multiway range trees: scalable IP lookup with fast updates, *Computer Networks* 44 (2004), 289–303.

Search Engines and Web Information Retrieval*

Alejandro López-Ortiz

School of Computer Science,
University of Waterloo,
Waterloo, Ontario, Canada
alopez-o@uwaterloo.ca

Abstract. This survey describes the main components of web information retrieval, with emphasis on the algorithmic aspects of web search engine research.

1 Introduction

The field of information retrieval itself has a long history predating web search engines and going back to the 1960s. From its very beginnings there was an algorithmic component to this field in the form of data structures and algorithms for text searching. For most of this time the state of the art in search algorithms stayed well ahead of the largest collection size. By the late 1980's with the computerization of the new edition of the Oxford English Dictionary the largest text collections could be searched in subsecond times. This situation prevailed until the second half of 1995 when the web reached a size that would tax even the best indexing algorithms. Initially the web was manually indexed via the "What's new?" NCSA web page, a role later taken over by Yahoo!. By early 1995 the growth of the web had reached such a ce that a comprehensive, manually maintained directory was no longer practicable. At around the same time researchers at Carnegie Mellon University launched the Lycos indexing service which created an index over selected words in web pages. In the Fall of 1995, OpenText started crawling the entire web and indexing "every word of every page". This last was the first comprehensive index of the Web. It utilized state of the art indexing algorithms for searching the Web. Shortly thereafter DEC launched the Altavista search engine which used what was then a massive computer to host their search engine. Since then there has been a steady stream of theoretical and algorithmic challenges in web information retrieval. Today Google indexes around eight billio pages using a cluster of an estimated size of 100,000 computers.

In this survey we consider the main challenges of web information retrieval. Typically a web search engine performs the following high level tasks in the process of indexing the web:

---

* A shorter version of this survey appears as part of "Algorithmic Foundations of the Internet", A. López-Ortiz, SIGACT News, v. 36, no. 2, June 2005.

A. López-Ortiz and A. Hamel (Eds.): CAAN 2004, LNCS 3405, pp. 183–191, 2005.

Typically a web search engine performs the following three high level tasks in the process of indexing the web:

1. Crawling, which is the process of obtaining a copy of every page in the web.
2. Indexing, in which the logical equivalent of the index at the back of a book is created.
3. Ranking, in which a relevance ordering of the documents is created.

Over the next few sections we will we discuss in more detail how to implement each of these steps.

2   Crawling

Discovery Process Search engines start by collecting a copy of the web through a process known as crawling. Starting from an arbitrary URL, a program called a "spider" (which "crawls the web") downloads the page, stores a copy of the HTML and identifies the links (anchor `<a href="..."`> tags) within the page. The "spider" is sometimes also called a "crawler" or a "robot".

The URL of each link encountered is then inserted into a database (if not already present) and marked as uncrawled. Once the spider has downloaded and completed the analysis of the current page it proceeds to the next uncrawled page in the database. This process continues until no new links have been found. In the past search engines only crawled pages that were believed to be "static" such as .html file as opposed to pages which were the result of queries (e.g. CGI scripts). With the increase of dynamically served content search engines now crawl what before would have been considered dynamic content. Nowadays a large commercial web site is usually served from a database, which has control over who can access and update the content and how often it must be refreshed. So in a certain sense, all the content appears to be dynamic. Crawlers must distinguish then this dynamically served, but rather static, content from a true dynamic web page such as a registration page.

If we create a graph in which each web page is a node and each HTML link is a directed edge, the crawling process will only reach those pages that are linked to starting from the arbitrary URL. To be more precise:

Definition 1. *The* web-page graph, *denoted as P, has a node for every web page and the directed edge (u,v) if web page u has an html link linking to page v.*

In these terms, the crawling process described above will succeed if and only if there is a path in the web graph from the starting node to all other nodes in the graph. In practice it has been observed that this is not the case. Indeed the web-page graph is formed of thousands of disconnected components. To make things more complicated the web-page graph is a directed graph as links are not bidirectional. That is to say, if page a links to page b, then b is reachable from a, but the converse is not true, unless b explictly links back to page a. This process needs to be repeated a large number of times each with a different starting URL.

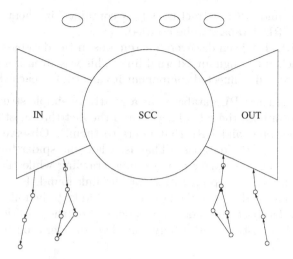

Fig. 1. Bow-tie structure

In fact Broder et al. [2] showed that the web-page graph has a bow-tie structure as shown in Figure 2. The center knot, or core is a strongly connected component (SCC) on the web-page graph. In this core we can reach any one page starting from any other page in the core. The left part of the tie consists of pages that point to the SCC core but are not themselves reachable from the core by "clicking" on links from page to page. These pages are called *origination* pages (IN).

The right part of the tie are those pages that are pointed to by at least one page in the core, but they themselves do not point back to the core. Those pages are called "termination" pages (OUT). The finger-like structures are called tendrils, which are connected to the main bow-tie but in the "wrong" direction (dead-ends). Lastly there are the disconnected components. At the time of the study the number of them was estimated at 17 million, however since then there has been some indication that their prevalence might have been overestimated due to the practice by some commercial web sites to return non-standard "404 Not Found" pages.

Observe that for the crawler to collect all pages, it must visit all tendrils and pages in the origination section. This means that the crawler has to be manually seeded to start millions of searches in the origination and disconnected section.

Clearly, these seeds cannot be discovered by following links alone. Hence alternative methods to collect URLs are needed. In practice some commonly used ones are: submit URL, random IP/URL probing, passive listening, and previously discovered URLs.

Certain disconnected components can still be reached if their creators manually submitted the URL of their web site to the search engine. A second method is for the crawler to perform random samplings by generating a random IP address not already in the URL database and requesting a web page from that address. Passive listening consists of scanning open communication channels such as newsgroups and IRCs extracting URLs contained in them, which are then added to the URL database to be crawled.

Once an URL has been discovered it remains in the database even if later on cannot be reached through an external link. This way even if a page eventually becomes part of a disconnected component it can still be reached by the crawler.

URL Database. The URL database has a relatively simple structure. The challenge comes from both the rate of access and the fact that most most web pages have more than one valid URL that refers to them.   Observe that URLs can be added off-line to the database. That is, when the spider finds a link in the process of crawling, the spider can continue crawling while the URL database processes the insert/lookup operation for the link found. Later on the crawler can request from the database the next batch of links to be followed. This means that the URL database does not need to depend on the traditional on-line data structures such as B-trees which fully complete an operation before proceeding on to a second one.

Document Duplication Detection. Another challenge is the duplicate document detection problem. Popular documents tend to be replicated numerous times over the network. If the document is in the public domain, users will make local copies available for their own benefit. Even if the popular content in question is proprietary, such as the web page of a news organization, it is likely to be duplicated (mirrored) by content distribution networks for caching purposes. Under certain circumstances a crawler can chance upon a mirror site and index this alternate copy.

In the case of public domain content, duplicate detection is complicated by the fact that users tend to change formatting and add extra headers and footers when creating a local copy. Hence we cannot use a straightforward tool such as checksum to check if two documents are identical. A second challenge in document duplication is the speed at which the documents must be processed. If time were not an issue, tools such as UNIX's `diff` whose performance is proven could be used to compare documents in a pairwise fashion. In the context of web page crawling this solution is not practicable as there are over four billions of pages on the web and hence a pairwise comparison would be too time consuming. To be effective and algorithm must make a determination of duplication within the time it takes to download a document.

Web Site Duplication Detection. Certain commercial web sites replicate their entire web sites in various geographic locations around the world. In such cases we would like to determine that the entire site is duplicated and hence it suffices to crawl only one of the copies. See [10, 3] which survey some of these problems in further detail.

3   Indexing

Once a copy of the web is available locally, the search engine can, in principle, receive queries from users searching for documents containing certain keywords or patterns. There are several different ways to accomplish this. For example, for small, rapidly changing files UNIX provides a facility called grep which finds patterns in a source text in time proportional to the length of the text. Clearly, in the case of the web which has an estimated size of several terabytes of text, grep would be too slow. Hence, we can benefit from the use of an index. A computerized index for a source text is not unlike the index of a book, where relevant words are listed together with the page number in which they appear. In the case of the web, we can generally index every word (relevant or not) appearing in the page or even every pattern together with the position in which they can be found in the corpus.

Definition 2. *The set of positions where a term can be located in the corpus is called the* postings set.

Typically, a user query is a collection of terms such as (`algorithm`, `index`, `internet`) which is interpreted as the set of documents containing all three terms. In other words, the answer to the query above is the intersection of the postings sets corresponding to each of the three terms.

Currently there are three main approaches to text indexing. The most extensively studied, from a theoretical perspective, is the *String Matching Problem* (SMP). In this problem the corpus is a single linear string of text. If there is more than one source document they are simply concatenated sequentially with some suitable markings to form one long string. For example, let the corpus consist of two documents as shown in Figure 3.

```
< d o c > N o t e s   I < / d o c > < d o c > N o t e s   I I < / d o c >
1 2 3 4 5 6 7 8 910 1 2 3 4 5 6 7 8 920 1 2 3 4 5 6 7 8 930 1 2 3 4 5 6 7
```

Fig. 2. Corpus with two documents. The numbers below denote the position of each letter in the text

The user queries for an arbitrary pattern, such as "otes". The corresponding output to the query is the position of all occurences of the pattern in the text. For example "otes" appears twice in the string above, with the first occurrence in position 7 and the second in position 25. The location is usually represented as a byte offset from the beginning of the corpus.

Another approach to search engines is the *Inverted Word Index*. In this case we only index words (or tokens) but not arbitrary patterns. Currently most, if not all, commercial web search engines use an inverted word index approach in their indices.

The third approach is the document based approach. As described above queries do not depend on the particular location of a word in a document, but only care about the presence or absence of the word in a given document. Indeed many search engines support this model only. Surprisingly, although widely used, this approach had not been formalized until recently.

3.1   The String Matching Problem

Suffix Trees. Traditionally algorithms supporting this type of queries use algorithms based on suffix trees. Suffix trees can be built in linear time using the algorithm of Esko Ukkonen [15]. In terms of space the straightforward implementation consumes $n \lg n$ bits of space. This is so as every tree edge is represented by a pointer of $\lg n$ bits long plus the position of the pattern in the text itself, stored in the leaves, also takes $\lg n$ bits. He et al. showed that the internal nodes can be implemented more succinctly using only a total of $O(n)$ space [9]. However the cost of the pointers to the text at the leaves remains the same for a total cost of $O(n \lg n)$.

For the case of the web n is in the order of trillions of bytes (2^{40}) and hence $\lg n \approx 40$. This means that a suffix based solution might require substantially more space than the original corpus size. Reducing this requirement is an active area of research and somewhat more efficient representations are known.

Burrows-Wheeler Transform. The Burrows-Wheeler transform can be used for more space efficient indexing techniques. The algorithm for searching for a pattern P using the BWT takes time $O(|P| \log n + occ)$ and is as follows.

In terms of text indexing, to date the biggest challenge has been to devise a search structure that supports searches for a keyword or pattern P on a text of n bits long in time $|P|$ (i.e. proportional the length of the pattern P) while using an index of size n (bit probe model). To place this in context, we know that searching can be done using $O(1)$ space, aside from the text, at the cost of search time $O(n)$ where n is the length of the text. This is the case for grep-like algorithms such as Knuth-Morris-Pratt. Such an algorithm of course would be impracticable for the web, hence the preference for SMP-like solutions that have $O(|P|)$ search time, or a close approximation, at the expense of $\Theta(n \log n)$ space usage.

The state of the art in indexing techniques is very close to achieving the dual objective of linear time i.e. $O(|P|)$ and linear space index structures. The classic suffix tree structure supports searches in $|P|$ time but requires an index of size $n \log n$. In contrast the recently developed compressed suffix arrays support searches in time $O(|P|/ \log n + \log^\epsilon n)$ using $O(n)$ bits in the unit cost RAM model [8, 12, 5].

3.2   Inverted Word Index

In this case we collect a set of words which are indexed using a standard index structure such as a sorted array, B-tree, or skip list. When the user queries for a word we search in the B-tree using the word as a key. This leads to an external leaf of the B-tree which points to the posting lists of the word, if present in the text.

In practice, an English text results in an index of size around 20% of the original size using naive methods, and as little as 5-10% using some advanced compressing mechanisms.

3.3   Document Index

Another algorithmic challenge of interest is to devise data structures and algorithms tailored to an heterogeneous, document-based collection of documents such as the World Wide Web [13, 4]. Most of the classic indexing schemes presume a model in which the user is searching for the specific location of a pattern in a contiguous text string. In contrast, search engine users give a collection of query terms (around three or so on average) and are searching for the subset of documents that contain some or all of the terms, ranked by some relevance metric. Observe that in this setting the input is a set of strings or documents each with a unique ID and there is no inherent order between different documents.

The operations required in the abstract data type (ADT) are

- $list(p)$ Report all documents containing the pattern p.
- $mine_k(p)$ Report all documents containing at least k occurrences of the pattern p, for a fixed, predetermined k.
- $repeats_k(p)$ Report all documents containing at least two occurrences of the pattern p at less than k positions away, for a fixed predetermined k.

S. Muthukrishnan gives algorithms for this problem [13]. A related problem consists of searching a large collection of documents that has been suitably tagged using XML. In such a setting the query language is extended to include predicates on the tags themselves, such as the XPath query language. There have been papers that study algorithms specifically tailored to XPath queries [6].

4   Ranking

Aside from the algorithmic challenges in indexing and query processing, there are ranking and classification problems that are, to a certain extent, unique to web content. Web publishing is not centrally managed; as a result, content is of varied quality and, as noted above, duplicates of popular documents are common. At the same time, and in contrast to other heterogeneous collections, content is crossed-linked by means of links (`<A HREF="`...`">`). In practice it has been observed that this structure can be exploited to derive information on the relevance, quality, and even content, of a document. To give a trivial example, the number of incoming links to a page is a reflection of the popularity of the topic as well as the quality of the document. In other words, all other things being equal, the higher the quality/usefulness of the document, the more links it has. In practice this is not a very effective method to determine document relevance as popularity of the topic seems to occlude relevance. For example, a very popular page that mentions an obscure term in passing will have more links that an authorative page defining and discussing the term.

In 1998, Jon Kleinberg at IBM, and independently Larry Page et al. at Stanford, discovered a way to distinguish links due to popularity from those reflecting quality [11, 14]. Links are initially given equal weights and considered as votes of confidence on other web sites. After this is done the links from each web site are re-weighted to reflect the confidence ranking computed above and the process is repeated. In principle this process could go on forever, never converging as sites transfer weights between them. A key observation was to consider first the induced subgraph of the result set from the web graph, and then interpret the adjacency matrix A of that subgraph as a linear transformation. Then one can show that the re-weighting process in fact converges to the eigenvalues of the matrix $A^T A$. The eigenvalues rank the pages by relative relevance, as perceived by their peers. This presumes that when a web site links to another there is, to a large degree, an implicit endorsement of the quality of the content. If this is the case often enough, the eigenvalue computation will produce an efficient ranking of the web pages based purely on structural information.

It is difficult to overestimate the importance of this result. Traditionally, many, if not most, systems for ranking results were based on natural language processing. Unfortunately, natural language processing has turned out to be a very difficult problem, thus making such ranking algorithms impractical. In contrast, the eigenvalue methods used by Page's ranking and Kleinberg's HITS (hubs and authorities) do not require any understanding of the text and apply equally to pages written in any language, so long as they are HTML tagged and mutually hyperlinked.

A drawback of the hub and authorities method is that the amount of computation required at query time is impractically high. An open problem is to find an alternate method to compute the same or a similar ranking in a more efficient manner.

Interestingly the same eigenvalue computation can be used to categorize the pages by topic as well as perform other content-based analysis [7]. This subfield has come to be known as spectral analysis of the web graph. [1] show that under a reasonable probabilistic model for the web graph, spectral analysis is robust under random noise scenarios.

5   Conclusions

In this survey we presented an overview of the main aspects of web information retrieval, namely crawling, indexing and ranking.

References

1. Yossi Azar, Amos Fiat, Anna R. Karlin, Frank McSherry, and Jared Saia. Spectral Analysis of Data. In Proceedings of *ACM Symposium on Theory of Computing* (STOC), 2001, pp. 6 19–626.
2. Andrei Broder, Ravi Kumar, Farzin Maghoul, Prabhakar Raghavan, Sridhar Rajagopalan, Raymie Stata, Andrew Tomkins and Janet Wiener. Graph structure in the Web, *Proceedings of the 9th international World Wide Web Conference*, 2000, pp. 309–320.

3. S. Chakrabarti, B. Dom, D. Gibson, J. Kleinberg, S.R. Kumar, P. Raghavan, S. Rajagopalan, and A. Tomkins. Mining the link structure of the World Wide Web. *IEEE Computer*, August 199 9.

4. Erik D. Demaine, Alejandro López-Ortiz, and J. Ian Munro. Adaptive set intersections, unions, and differences. In Proceedings of *ACM-SIAM Symposium on Discrete Algorithms* (SODA) 2000, pp. 743-752.

5. Erik D. Demaine, and Alejandro López-Ortiz. A linear lower bound on index size for text retrieval. *Journal of Algorithms*, Vol. 48, no. 1, pp. 2-15, 2003.

6. Richard F. Geary, Rajeev Raman, and Venkatesh Raman. Succinct ordinal trees with level-ancestor queries. In Proceedings of *ACM-SIAM Symposium on Discrete Algorithms*, (SODA) 2004, pp. 1–10.

7. D. Gibson, J. Kleinberg, and P. Raghavan. Inferring web communities from link topology. In Proceedings of the *ACM Conference on Hypertext and Hypermedia*, 1998, pp. 225–234.

8. Roberto Grossi, and Jeffrey Scott Vitter. Compressed suffix arrays and suffix trees with applications to text indexing and string matching. In Proceedings of *ACM Symposium on Theory of Computing* (STOC), 1999, pp. 3 97–406.

9. Meng He, J. Ian Munro and S. Srinivasa Rao. A Categorization Theorem on Suffix Arrays with Applications to Space-efficient Text Indexes. To appear in proceedings of *ACM-SIAM Symposium on the Discrete Algorithms*, (SODA), 2005.

10. Monika R. Henzinger. Algorithmic Challenges in Web Search Engines. *Internet Mathematics*, vol. 1, no. 1, 2004, pp. 115-126.

11. J. Kleinberg. Authoritative sources in a hyperlinked environment. In Proceedings of *ACM-SIAM Symposium on Discrete Algorithms* (SODA), 1998, pp. 668–677.

12. U. Manber, and G. Myers. Suffix arrays: a new method for on-line string searches. *SIAM Journal on Computing*, vol. 22, no. 5, 1993, pp. 935–948.

13. S. Muthukrishnan. Efficient algorithms for document retrieval problems. In Proceedings of *ACM-SIAM Symposium on Discrete Algorithms*, (SODA) 2002.

14. Lawrence Page, Sergey Brin, Rajeev Motwani, and Terry Winograd. The PageRank Citation Ranking: Bringing Order to the Web. *Technical Report, Department of Computer Science, Stanford University*, 1999-66.

15. Esko Ukkonen. On-Line Construction of Suffix Trees. *Algorithmica* v.14 n.3, pp.249-260, 1995.

Algorithmic Foundations of the Internet: Roundup*

Alejandro López-Ortiz

School of Computer Science,
University of Waterloo,
Waterloo, ON, Canada
alopez-o@uwaterloo.ca

Abstract. In this paper we present a short overview of selected topics in the field of Algorithmic Foundations of the Internet, which is a new area within theoretical computer science.

1   Structure of the Roundup

This paper presents a roundup of subjects not covered in the comprehensive surveys before it. These topics are also representative of the field of algorithmic foundations of the Internet. For each topic covered in the survey we start by describing at a high level what the problem or object of study is. We then give a brief historical perspective of how the challenge posed has been addressed so far within the network community, whenever relevant. Lastly we describe the theoretical and algorithmic aspects of the topic or challenge in question.

2   Web Caching

The world wide web was created in 1991, and became wildly popular in 1993 after the release of the graphics friendly Mosaic browser. Shortly thereafter it became clear that HTTP transactions would come to be the dominant consumer of bandwidth on the Internet [38]. Indeed this became the case sometime around 1995 and remained so until 2002, when peer-to-peer music and video sharing overtook HTTP traffic by volume. Even today it remains the second largest user of bandwidth, by protocol.

Being a large consumer of bandwidth means that reductions in unnecessary web retransmissions could potentially have a significant impact on the bandwidth requirements of an organization.

The term caching denotes the management of data being swapped between various storage media with differing transfer rates, and in particular, most often it is used to describe the transfer of data files or virtual memory between

---

* An earlier version of this survey appearead in "Algorithmic Foundations of the Internet", A. López-Ortiz, SIGACT News, v. 36, no. 2, June 2005.

A. López-Ortiz and A. Hamel (Eds.): CAAN 2004, LNCS 3405, pp. 192–204, 2005.

RAM and hard disks. For example, in this classical sense the operating system is responsible for paging-in the working set of a program, while the rest of the instruction data might reside in disk. Accessing data across the Internet, such as a web page, also involves differential rate accesses as in the classical setting. Hence it is only natural that web traffic is amenable to caching speedups. Surprisingly, the original data transfer protocols such as FTP and HTTP did not initially incorporate notions of caching in their operations. Yet the need for them is the same, if not greater.

In Figure 1 we illustrate the typical path of a web page request. Naturally, caching might occur at any node along the path. Furthermore, if we take advantage of some HTML features it is possible to cache data off-path, using a cache farm.

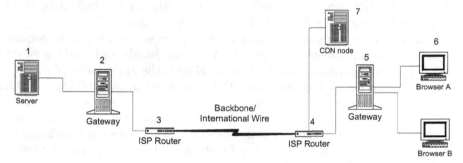

Fig. 1. Web page access path over the network

2.1   Web Caching in Practice

Efforts to develop a web caching infrastructure started as soon as it became clear that web traffic growth was unprecedented. The first organizational proxy cache was introduced by DEC in 1994. A proxy cache serves as a centralized store for all HTTP requests generated from within an organization. The proxy mediates between the faster internal LAN and the slower external Internet connection. This corresponds to node 5 in Figure 1. In this way if client A requests a web page and client B issues a second request for the same web page within a short period of time, the second request can be served from the proxy storage without causing further traffic on the external network.

Similarly, the Netscape browser introduced a hard-drive mapped file cache (see node 6 in Figure 1). This cache, which is now standard in all browsers, was designed with the primary consideration of speeding up the "back" operation in the browser. As the user backtracks on a sequence of links a local copy of a previously viewed file is served instead of requesting a new copy across the network. As trivial as this might seem such a local cache system was lacking in the Mosaic browser as well as in most access clients for the other two major file transfer protocols at the time, namely, FTP and gopher.

A national cache hierarchy was implemented in New Zealand in 1994. This corresponds to node 4 in Figure 1. A cache hierarchy is a caching infrastructure serving more than one organization or ISP (Internet Service Provider). Since the cost of international traffic over an undersea cable to New Zealand was high, great savings could be derived from pooling resources to create a national cache hierarchy.

Other forms of caching were slower in coming. Network-edge proxies, which cache content from the server to the network edge, (corresponding to node 2 in Figure 1) first appeared in the year 2000 or so. This type of caching reduces traffic on the internal network where the server is located. Since the internal network is typically used for business critical functions, while web servers rarely are, a network edge cache realigns the use of internal bandwidth with local parameters. Secondly, a network edge cache precludes external HTTP packets from traversing in the internal organizational network. Such external packets could potentially become a security risk, thus reducing or eliminating them is considered a desirable side effect of network edge caching.

Similarly geographic push caching was first proposed in [18], and subsequently supported by experimental setups [31]. Yet, surprisingly such caching schemes did not become popular until the year 2000 with the appearance of widely deployed content distribution networks (CDN) such as Akamai and Digital Island. In this type of caching the server proactively disseminates content to caching nodes located across the network. When the client requests a web page, its request is redirected to a geographically close cache server by a combination of DNS redirection and secondary content redirection, that is, redirection of <IMG...> links (see node 7 in Figure 1).

To this date, we are not aware of a caching solution that is deployed at the national level on the server side (node 3 in Figure 1). Similarly we do not know of any web specific caching algorithms at the web server: any caching that takes place in it is derived from the file cache management in the operating system.

2.2   Algorithmic and Theoretical Aspects

In terms of content networks, theoretical efforts have taken place to create optimal schemes for load balancing and content redirection in CDNs [27]. However these issues are secondary parameters in the efficiency of CDNs, as content location is preeminently mandated by geographic considerations rather than load balancing issues.

The problem of determining optimal geographic location for cache placement has been the subject of intense theoretical study. Consider first the extension of Figure 1 to include all nodes that access a specific web page, e.g., all users accessing the SIGACT home page as shown in Figure 2. The result is a graph with the SIGACT server as root and the computers running web browsers (clients) as nodes of degree 1 (see Figure 2). In general such a graph might have an arbitrary topology although it has been observed in practice that the graph obtained is very nearly a tree, in the sense that it has n nodes and $n+c$ edges, for a constant $c \ll n$. Consider now a CDN company which wishes to place k-caching servers on the tree nodes in such a way as to minimize the total amount of traffic used in serving the requests of the clients. The working assumption is that a

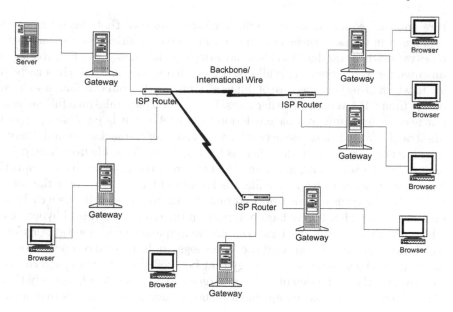

Fig. 2. Network of distribution paths for a single server

request is served by the first cache node appearing in the path from the client to the server. Computing the optimal location of the caching servers has been shown to be NP-complete by [28] for arbitrary topologies of the graph, but it is known to be computable for tree topologies [30]. Other variants of this problem have been considered, with differing assumptions of the model. In general, this problem is closely related to the well known p-median server problem, which is NP-complete. Hence many of the variants studied remain NP-complete under a general network topology or even assuming a more restricted Internet-like topology (see e.g. [24]).

Another important parameter in a CDN is the cache management strategy. That is, which files are replicated where and how are they evicted. Each server in a farm can be treated as an on-line isolated object store in which case the server may use classical standard caching strategies such as LRU-like policies (more on this later), or the entire set of k servers can act in a coordinated way, establishing a cache policy that also reflects centrally collected historical access patterns. These globally coordinated strategies are known as *replica placement algorithms* [24]. A general survey of the different algorithms can be found in [26] and an intriguing argument questioning the benefits of a coordinated policy over single server strategies appears in [25].

Web cache management strategies were first considered by Markatos who observed that web files are served on an all-or-nothing basis, since in most cases users either request the entire page or nothing at all [34]. This is in contrast to the classical setting, where, for example, access to data using a database does not result in loading the entire database file, but only the relevant portions of the index structure and the requested datum itself. Markatos observed that this difference was significant in terms of cache performance and gave empirical data to support the observation that standard RAM-to-hard-drive cache management strategies were unsuited for web servers. This all-or-nothing characteristic results in a cache policy in which objects cached are of different sizes. Aside from this fact, a web cache differs from a RAM cache in other ways. First, a web page miss on a full cache does not necessarily result in a page eviction from cache as it is possible to serve the file straight from disk to the network card bypassing the cache. Second, there is a dual latency parameter in that there is lag time to load the file from hard drive to memory and a second lag in serving the file across the network. This means that in practice when serving a large file, the latency of transmission over the network dominates the transfer time from external storage to memory. Hence a cache fault on a large file on hard drive has no impact on the latency perceived by web users. Third, the web server has access to extensive frequency of access statistics for the web files in question. In this way, the access logs can be used to compute an off-line or on-line cache management strategy that best reflects the access patterns to the web site. Lastly as the web objects are larger and accessed less frequently than in OS caching, it is possible to implement cache management strategies that are more computationally expensive per given access.

Irani was the first to propose and analyze an algorithm specifically designed for web files that took in consideration both frequency of access and multi-size pages [20, 9]. Interestingly, it was observed that the off-line optimum strategy of always evicting the page whose next request is furthest in the future—also known as Belady's rule—does not hold for multi-sized pages. Indeed, the offline optimum is NP-hard to compute under most models considered [4]. Another consequence of Belady's rule not holding is that the standard LRU-based on-line strategies used in the classic setting are not optimal for multi-sized pages. Indeed, the best known strategy for this problem has competitive ratio of $\Theta(\log n)$ in contrast to the constant competitive ratio of LRU in the classical server setting [20]. Several refinements of Irani's model have been proposed in an attempt to better reflect the costs of a cache miss in the web setting. Another aspect that has been incorporated is the second order correlations in the access patterns of web pages [11, 17]. Further details on web caching can be found in [21].

Aside from the benefits of reduced latency for the user (if there is a cache hit), web caching also results in increased processing capacity for the server (if the file can be served from the server's main memory), reduced server load (if the file can be served from a remote cache) and reduced demands on network bandwidth. Interestingly, this last not only benefits the client involved in the page request, but also other clients which will now encounter reduced network contention as a result of bandwidth savings from other users.

3   Internet Economics and Game Theory

The Internet has a decentralized structure with no overarching commanding authority. At the same time, this is not to say that anarchy reigns on the Internet.

To the contrary, the internet protocols fully specify the context and rules under which network interactions are to take place. On the other hand, if a specific feasible solution is not explicitly predicated by the protocols, there is no central authority to impose it. For example, consider a simple network as illustrated in Figure 3. In this case node A wishes to send a unit size message to node B while node C wishes to send a message to node D. We assume the capacity of all edges to be the same. A coordinating mechanism is necessary to ensure that only one of A or C chooses to send the message through edge e_1, while the other should choose edge e_2. The Internet does not provide a mechanism for this type of centralized coordinating action.

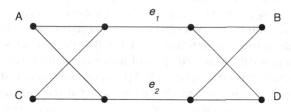

Fig. 3. Simple network

In practice what happens is that each organization in the network makes its choices separately and mostly independently. The quality of each user choice, however, is dependent on the choices of all other users. Hence it is to one's own benefit to consider what are the likely choices of other organizations and incorporate those into one's own reasoning. This is exactly the type of reasoning that takes place in a classical game setting, such as chess, monopoly or poker: there is a set of commonly agreed upon rules, no central coordinating authority and self-interest is the main motivator.

A natural question is what game is defined by the the current Internet. That is, what are the individual payoff functions for players, where do random events take place and what exactly are the rules of the game.

Alternatively, the converse question holds. Consider a given strategy for selecting state, such as the TCP/IP congestion control protocol. What is the payoff function that is maximized by this protocol as currently defined? It could well be the case that the current function being optimized is not desirable to anyone in particular in which case a change in the protocol would be called for.

To illustrate further, consider a simple Internet game in which the players are web servers, serving content across the network. Making a move consists of choosing an injection rate of packets into the network. If the injection rate chosen is too high, then some of the packets sent are destroyed en-route to the destination and must be retransmitted, thus in the end reducing the effective transmission rate. If on the contrary the packet injection rate is too low, the server could increase its utility function value by increasing its injection rate. Observe that the injection rate that can be sustained is dependent on the amount of traffic being sent over any given link by all players. The goal for each player is to maximize its own effective transmission rate. Clearly this defines a game in the classical game theory sense.

In this injection game, the players aim to adjust their injection rate until it cannot be improved any further, that is, if any one of them were to increase or decrease its current injection rate this would result in a lower effective transmission rate for that player. That is to say, the players aim to reach a Nash equilibrium point. This is defined as a game state in which no single player has an incentive to alter its current choice of play. Formally, an n player game is a finite set of actions S_i for player i and payoff or utility functions $u_i : S_1 \times S_2 \times \ldots \times S_n \to \mathcal{R}$ for $i = 1 \ldots n$. A *(pure) Nash equilibrium* is a state of the game $(s_1, \ldots, s_n) \in S_1 \times S_2 \times \ldots \times S_n$ such that $u_i(s_1, \ldots, s_i, \ldots, s_n) \geq u_i(s_1, \ldots, s'_i, \ldots, s_n)$ for all $i = 1 \ldots n$.

Currently the Internet uses a certain set of mechanisms (some by virtue of the protocol, others by common practices) to deal with injection rates that are too high. These mechanisms define the "rules" and hence the valid actions in the injection game above. A natural question to ask is if the rules as defined lead to optimal usage of the network. For example, it might be the case that the rules are overly cautious (e.g. TCP window size) and lead the players to select injection rates that are below what could be sustained over the current network topology. A formal analysis of the game should lead to the expected transmission rate that the players will reach. In turn this value can be compared with the maximum capacity of the network, as determined by standard network flow analysis.

A no less important question is if there exists *any* enforceable set of rules that could lead to an optimal usage of the network as deployed. In general it is known that there are games in which the Nash equilibrium points give suboptimal global behaviour, that is, in certain games the Nash equilibrium points result in outcomes undesirable to all parties (e.g. prisoner's dilemma, tragedy of the commons). These tend to arise in situations in which in the globally "best" solution there exists at least one player who can improve its lot by acting selfishly. The other players then must assume beforehand that such player will defect if given the opportunity and hence avoid such (otherwise preferable) states altogether. In other words, players actively avoid unstable states and aim for a Nash equilibrium in which players have no incentive to defect.

In this case it is important to determine how much worse is the outcome in the uncoordinated game setting as compared to a centrally mandated solution. For example, in the tragedy of the commons game scenario, there exists a shared grass meadow which is used for grazing. The globally optimal solution is for each of the n players to consume $1/n$ of the meadow at a sustainable rate. However, in such situation, a selfish player that consumes more than its fair share improves its utility function and hence has an incentive to cheat. This undesirable outcome can be avoided if we allow a central authority (typically the State, but not always necessarily so) to enforce a maximum rate of consumption of $1/n$. In the absence of such authority the optimal Nash equilibrium strategy in this case is for each player to consume as much of the meadow grass as possible, as fast as possible, until only the bare land is left (as actually happened in practice). In this last scenario all players end up being worse off.

The difference between the benefit players would have derived in the centrally planned solution and the Nash equilibrium solution is sometimes referred to as *the price of anarchy*.[1] For example, consider a routing game which is in some ways similar to the injection game. The players are nodes in the network, which can select one of several routes to send a message. The utility function for all players is to minimize the latency (travel time) of their individual messages. As in the injection game, the effective latency is affected by which routes are chosen by the other players. This game has a Nash equilibrium point as solution which we can compare with the globally optimal solution. The globally optimal solution is defined as that which maximizes the social welfare. In the routing game the social welfare is defined as the inverse of the sum of all latencies and the globally optimal solution is hence that which minimizes this sum of latencies[2].

Interestingly, it has been shown that the price of anarchy for the routing game is at most two in the worst case [37]. This is generally considered an acceptable penalty in exchange for the benefits of a globally distributed solution with no single point of failure and no need for active enforcement as players have no incentive to cheat.

Conversely, we could aim to modify the rules in such a way that in the resulting game the Nash equilibrium solution coincides with the globally optimal solution. This is sometimes refered to as *inverse game theory*. In standard game theory we are given a game and the goal is to identify the Nash equilibrium points. In inverse game theory we are given the desired state that we wish to achieve, and we wish to design a game that has that desired state as a Nash equilibrium point and no other. See [42] for a more detailed introduction to the subject of Internet routing and game theory.

4   Tomography

Efficient routing and caching require accurate connectivity information of the Internet. Internet protocols, on the other hand, make this task difficult as routing decisions are made locally and shared across organizations only in aggregate form.

During the early years of the Internet this did not pose much of a challenge as the backbone was administered by the NSF with a relatively open policy. However, since the NSF relinquished control of the Internet backbone in the

---

[1] Be warned that there is a bit of hyperbole in the term "anarchy" as used in this instance, as in all cases we assume that somehow the rules of the game are still being enforced. Indeed economists refer to this milder "anarchy" under consideration as *laissez faire*. The true anarchist, no-rules-barred optimal solution for a player in the injection game would be to unplug the other servers.

[2] Observe that the choice of social welfare is generally not unique. For example we could define the social welfare as the maximum over all latencies or the difference between the maximum and minimum latencies.

early 1990's the Internet has not been centrally managed. As a consequence, the topology of the network became a black box. In fact, backbone operators (National Service Providers or NSPs for short) generally consider the low level details of their respective backbone configurations a commercial secret both for competitive and security reasons.

Hence the information of interest, be it latency, topology, or connectivity, has to be inferred from experimental measurements. The inference of network topology and other network properties through indirect measurement came to be known as *internet tomography* [13, 14, 41].

4.1   Tomography in Practice

Internet protocols are generally designed to build a layer of abstraction above the underlying hardware. This is a great advantage in terms of interoperability: since not much is assumed from the physical network media, most network hardware protocols can be (and have been) seamlessly incorporated into the Internet. The flip side of this is that the layer of abstraction hides low level details which might be useful for measuring certain network parameters such as routing quality, stability and security. Fortunately, not all low level details are hidden. In particular two of the most commonly used tools for low level discovery are `ping` and `traceroute` or derivatives of them. `ping` allows a computer to measure the round trip time for a message on between two computers on the Internet. `traceroute` returns the path taken by a packet from source to destination. Another effective method for obtaining connectivity information is to examine a set of BGP routing tables on the Internet. The routing table at a given node on the Internet contains a high level view of the routes used from that point on to any other computer on the network.

Paxson initiated the study of network properties using data collected through external access points [35]. Currently there are several efforts in progress to obtain topology and performance measurements on the Internet, several of which use some form of measurement points or agents called *beacons*. Beacons inject traffic into the network to extract information from the reported path and latency of the traffic generated. In practice these measurement points (beacons) are often placed in universities and other organizations that are willing to host the software or hardware required. For example the National Internet Measurement Infrastructure (NIMI) [2, 36, 16] is a concerted effort to deploy general purpose beacons with particular focus in scalability and flexibility. Building on top of the NIMI effort is the Multicast-based Inference of Network-internal Characteristics (MINC) project which aims to measure performance characteristics of the network through the use of end-to-end measurements [8]. Other efforts along these lines are [23, 40, 41, 39, 16]. The location of these beacons is determined according to various heuristics [1, 3, 6, 12].

4.2   Algorithmic and Theoretical Challenges in Tomography

As we have seen, there have been substantial efforts directed at the deployment and use of distributed measurement systems. A key question is, what are the properties necessary for such a system to provide accurate measurements? At least in principle, there could be properties that cannot be measured from end-to-end beacons or the measurements returned by the beacons could be skewed in some manner. This is not a purely academic concern: Chen et al. showed that in the case of the high level AS topology, some of the techniques in use provided biased samples [10].

To be more precise, the Internet is divided into high level organizations called Autonomous Systems (AS). Each of these autonomous systems is a grouping of thousands of computers. Any two given ASes have few to no direct links in between them. In this way one can create a high level map of the topology of the network by considering one node per AS and direct connections between them as edges. Some early measurements of the AS topology were derived from routing tables—a seemingly sensible measurement technique. Such measurements suggested a power law distribution of the node degrees in the AS map [15]. The study by Chen et al. measured the same topology using an alternate, more accurate technique and obtained an AS topology map of the Internet which did not evidence a power law degree distribution [10, 32, 33]. In principle the same could possibly be the case for end-to-end measurements derived from beacons. This question can be addressed both at the empirical and theoretical levels, as we shall see.

Jamin et al. proposed theoretical methods as well as ad hoc heuristics for computing the location of a set of beacons whose aim is to compute the distance maps on the network [22]. Recently, Barford et al. provided the first systematic experimental study to validate the empirical observation that a relatively small number of beacons is generally sufficient to obtain an accurate map of the network [5]. Moreover, they show that the marginal utility of adding active measurement points decreases rapidly with each node added. They also provide an intriguing theoretical model of the marginal utility of an additional repeated experiment in a sequence of statistical measurements. Bu et al. consider the problem of the effectiveness of tomography on networks of general topology [7]. While their focus is on the ability to infer performance data from a set of multicast trees, their systematic study of the abilities of tomography in arbitrary networks has strong theoretical underpinnings.

Horton et al. showed that determining the number and location of the minimum beacon set is NP-hard for general topologies and in fact it is not even approximable to a factor better than $\Omega(\log n)$ [19]. Worse still in some networks the minimum number of beacons needed is $(n - 1)/3$. If we consider that the Internet has on the order of 285 million computers, a 95 million computer beacon set would be impractically large. Interestingly, the theoretical analysis suggested reasons why a much smaller beacon set would suffice on the current Internet topology. The paper gives an effective method to bound the number of beacons needed—at somewhere less than 20,000 nodes. Building upon these results Kumar et al. propose a refinement of the model which produces a robust beacon set of even smaller size [29].

5   Conclusions

The field of algorithmic foundations of the Internet has seen rapid growth over the last decade. Results in algorithmic foundations of the internet regularly appear in general theory conferences. At the same time there has also been an increase in the number of papers with a theoretical bent in what traditionally had been applied networks conferences. This new field has attracted mathematicians, physicists, combinatorists, management science/economists, and, naturally, computer scientists. The challenges are numerous and developments are often immediate applicability to the internet.

References

1. A. Adams, T. Bu, R. Caceres, N. Duffield, T. Friedman, J. Horowitz, F. Lo Presti, S.B. Moon, V. Paxson, and D. Towsley. The Use of End-to-end Multicast Measurements for Characterizing Internal Network Behavior, *IEEE Communications*, May 2000.
2. A. Adams, J. Mahdavi, M. Mathis, and V. Paxson, Creating a Scalable Architecture for Internet Measurement. *Proceedings of the 8th Internet Society Conference (INET)*, 1998.
3. M. Adler, T. Bu, R. K. Sitaraman, and D. F. Towsley. Tree Layout for Internal Network Characterizations in Multicast Networks. *Networked Group Communication*, 2001, pp. 189-204.
4. Susanne Albers, Sanjeev Arora, and Sanjeev Khanna. Page replacement for general caching problems. In Proceedings of *ACM-SIAM Symposium on Discrete Algorithms*, (SODA) 1999, pp. 31–40.
5. P. Barford, A. Bestavros, J. W. Byers, and M. Crovella. On the marginal utility of network topology measurements. *Internet Measurement Workshop*, 2001, pp. 5-17.
6. S. Branigan, H. Burch, B. Cheswick, and F. Wojcik. What Can You Do with Traceroute? *Internet Computing*, vol. 5, no. 5, 2001, page 96ff.
7. T. Bu, N. G. Duffield, F. Lo Presti, and D. F. Towsley. Network tomography on general topologies. *ACM Conference on Measurements and Modeling of Computer Systems (SIGMETRICS)* 2002, pp. 21-30
8. R. Caceres, N.G. Duffield, J. Horowitz, and D. Towsley. Multicast-based inference of network internal loss characteristics. *IEEE Transactions on Information Theory*, v.45, n.7, 1999, pp. 2462-2480.
9. Pei Cao, and Sandy Irani. Cost-Aware WWW Proxy Caching Algorithms. In Proceedings of *USENIX Symposium on Internet Technologies and Systems*, 1997.
10. Qian Chen, Hyunseok Chang, Ramesh Govindan, Sugih Jamin, Scott Shenker, and Walter Willinger. The Origin of Power-Laws in Internet Topologies Revisited. In Proceedings of *IEEE Conference on Computer Communications*, INFOCOM, 2002.
11. Ludmila Cherkasova, and Gianfranco Ciardo. Role of Aging, Frequency, and Size in Web Cache Replacement Policies. In Proceedings of *9th International Conference High-Performance Computing and Networking*, (HPCN Europe) 2001, pp. 114–123.

12. Bill Cheswick, Hal Burch, and Steve Branigan. Mapping and Visualizing the Internet. *Proc. USENIX Technical Conference*, 2000.
13. K. Claffy, G. Miller and K. Thompson. The nature of the beast: recent traffic measurements from an Internet backbone. *Proc. 8th Internet Soc. Conf. (INET)*, 1998.
14. K. Claffy, T.E. Monk and D. McRobb. Internet Tomography. *Nature*, 7th January 1999.
15. Michalis Faloutsos, Petros Faloutsos, Christos Faloutsos. On Power-law Relationships of the Internet Topology. In Proceedings of *ACM Conference on Applications, Technologies Architectures and Protocols for Computer Communication* (SIGCOMM), 1999. pp. 251–262.
16. P. Francis, S. Jamin, V. Paxson, L. Zhang, D. F. Gryniewicz, Y. Jin. An Architecture for a Global Internet Host Distance Estimation Service. *Proceedings of the IEEE Conference on Computer Communications (INFOCOM)*, 1999, pp. 210-217
17. Alexander Golynski, Alejandro López-Ortiz, and Ray Sweidan. Exploiting Statistics of Web Traces to Improve Caching Algorithms. *Technical Report CS-2003-34, School of Computer Science, University of Waterloo*, 2003.
18. J. Gwertzman, and M. Seltzer. The Case for Geographical Push-Caching, in VINO: The 1994 Fall Harvest, *Technical Report TR-34-94, Center for Research in Computing Technology, Harvard University*, December, 1994.
19. Joseph D. Horton, and Alejandro López-Ortiz. On the Number of Distributed Measurement Points for Network Tomography, In Proceedings of *ACM Internet Measurements Conference* (IMC), pp. 204-209, 2003.
20. Sandy Irani. Page Replacement with Multi-Size Pages and Applications to Web Caching. In Proceeedings of *ACM Symposium on the Theory of Computing* (STOC) 1997, pp. 701–710.
21. S. Irani. Competitive Analysis of Paging. *Online Algorithms*, Lecture Notes in Computer Science 1441, Springer, 1998, pp.52-71.
22. S. Jamin, C. Jin, Y. Jin, D. Raz, Y. Shavitt, L. Zhang. On the Placement of Internet Instrumentation. *IEEE Conference on Computer Communications (INFOCOM)*, 2000, pp. 295-304.
23. S. Kalidindi and M. J. Zekauskas. Surveyor: An infrastructure for Internet performance measurements. *Proceedings of the 9th Internet Society Conference (INET)*, ISOC, 1999.
24. Jussi Kangasharju, James W. Roberts, Keith W. Ross. Object replication strategies in content distribution networks. *Computer Communications*, 2002, vol. 25, no.4, pp. 376–383.
25. Magnus Karlsson and Mallik Mahalingam. Do We Need Replica Placement Algorithms in Content Delivery Networks. In Proceedings of *7th International Workshop on Web Content Caching and Distribution* (WCW), 2002.
26. Magnus Karlsson, Christos Karamanolis, and Mallik Mahalingam. *A unified framework for evaluating replica placement algorithms*. Technical Report HPL-2002, HP Laboratories, July 2002.
27. David R. Karger, Eric Lehman, Frank Thomson Leighton, Rina Panigrahy, Matthew S. Levine, Daniel Lewin. Consistent Hashing and Random Trees: Distributed Caching Protocols for Relieving Hot Spots on the World Wide Web. Proceedings of *ACM Symposium on the Theory of Computing* (STOC) 1997, pp.654–663.
28. P. Krishnan, Danny Raz, and Yuval Shavitt. The cache location problem. *IEEE Transactions on Networking*, vol. 8, no. 5, 2000, pp. 568–582.

29. Ritesh Kumar, and Jasleen Kaur. Efficient Beacon Placement for Network Tomography. In Proceedings of *ACM Internet Measurements Conference* (IMC), 2004.

30. Bo Li, Mordecai J. Golin, Giuseppe F. Italiano, Xin Deng and Kazem Sohraby. On the optimal placement of web proxies in the internet. Proceedings of *IEEE Conference on Computer Communications* (INFOCOM) 1999, pp.1282–1290.

31. A. López-Ortiz and D. M. Germán, A Multicollaborative Push-Caching HTTP Protocol for the WWW, *Poster proceedings of the 5th International World Wide Web Conference* (WWW96), 1996.

32. Zhuoqing Morley Mao, Jennifer Rexford, Jia Wang, and Randy H. Katz. Towards an accurate AS-level traceroute tool. In Proceedings of the *ACM Conference on Applications, technologies, architectures, and protocols for computer communications* (SIGCOMM), 2003.

33. Zhuoqing Morley Mao, David Johnson, Jennifer Rexford, Jia Wang, and Randy Katz. Scalable and Accurate Identification of AS-Level Forwarding Paths. Proceedings of *IEEE Conference on Computer Communications* (INFOCOM), 2004.

34. Evangelos P. Markatos. Main Memory Caching of Web Documents. Proceedings of the *Fifth International World Wide Web Conference* (WWW96), 1996.

35. V. Paxson. Measurements and Analysis of End-to-End Internet Dynamics. *PhD thesis, Univ. of Cal., Berkeley,* 1997.

36. V. Paxson, J. Mahdavi, A. Adams and M. Mathis, An Architecture for Large-Scale Internet Measurement. *IEEE Communications,* v.36, n.8, 1998, pp. 48–54.

37. Tim Roughgarden, and Éva Tardos. How Bad is Selfish Routing?. In Proceedings of *IEEE Symposium on Foundations of Computer Science* (FOCS) 2000, pp. 93–102.

38. J. Sedayao. Mosaic Will Kill My Network! Studying Network Traffic. Proceedings of the *First International Conference on the World-Wide Web* (WWW94), Elsevier, May, 1994.

39. R. Siamwalla, R. Sharma, and S. Keshav. Discovering Internet Topology. Technical Report, Cornell University, July 1998.

40. Cooperative Association for Internet Data Analysis (CAIDA). *The Skitter Project.* http://www.caida.org/tools/measurement/skitter/index.html, 2001.

41. D. Towsley. Network tomography through to end-to-end measurements. Abstract in *Proc. 3rd Workshop on Algorithm Engineering and Experiments (ALENEX),* 2001.

42. Berthold Vöcking. Selfish Routing and Congestion Games: Towards a game based analysis of the Internet. *School of Algorithmic Aspects of Game Theory in Large Scale Networks,* Santorini, Greece, 2004. http://www.cti.gr/AAGTLSN/#LECTURERS.

Author Index

Lecture Notes in Computer Science

For information about Vols. 1–3492

please contact your bookseller or Springer

Vol. 3540: H. Kalviainen, J. Parkkinen, A. Kaarna (Eds.), Image Analysis. XXII, 1270 pages. 2005.

Vol. 3537: A. Apostolico, M. Crochemore, K. Park (Eds.), Combinatorial Pattern Matching. XI, 444 pages. 2005.

Vol. 3536: G. Ciardo, P. Darondeau (Eds.), Applications and Theory of Petri Nets 2005. XI, 470 pages. 2005.

Vol. 3535: M. Steffen, G. Zavattaro (Eds.), Formal Methods for Open Object-Based Distributed Systems. X, 323 pages. 2005.

Vol. 3533: M. Ali, F. Esposito (Eds.), Innovations in Applied Artificial Intelligence. XX, 858 pages. 2005. (Subseries LNAI).

Vol. 3532: A. Gómez-Pérez, J. Euzenat (Eds.), The Semantic Web: Research and Applications. XV, 728 pages. 2005.

Vol. 3531: J. Ioannidis, A. Keromytis, M. Yung (Eds.), Applied Cryptography and Network Security. XI, 530 pages. 2005.

Vol. 3530: A. Prinz, R. Reed, J. Reed (Eds.), SDL 2005: Model Driven. XI, 361 pages. 2005.

Vol. 3528: P.S. Szczepaniak, J. Kacprzyk, A. Niewiadomski (Eds.), Advances in Web Intelligence. XVII, 513 pages. 2005. (Subseries LNAI).

Vol. 3527: R. Morrison, F. Oquendo (Eds.), Software Architecture. XII, 263 pages. 2005.

Vol. 3526: S.B. Cooper, B. Löwe, L. Torenvliet (Eds.), New Computational Paradigms. XVII, 574 pages. 2005.

Vol. 3525: A.E. Abdallah, C.B. Jones, J.W. Sanders (Eds.), Communicating Sequential Processes. XIV, 321 pages. 2005.

Vol. 3524: R. Barták, M. Milano (Eds.), Integration of AI and OR Techniques in Constraint Programming for Combinatorial Optimization Problems. XI, 320 pages. 2005.

Vol. 3523: J.S. Marques, N. Pérez de la Blanca, P. Pina (Eds.), Pattern Recognition and Image Analysis, Part II. XXVI, 733 pages. 2005.

Vol. 3522: J.S. Marques, N. Pérez de la Blanca, P. Pina (Eds.), Pattern Recognition and Image Analysis, Part I. XXVI, 703 pages. 2005.

Vol. 3521: N. Megiddo, Y. Xu, B. Zhu (Eds.), Algorithmic Applications in Management. XIII, 484 pages. 2005.

Vol. 3520: O. Pastor, J. Falcão e Cunha (Eds.), Advanced Information Systems Engineering. XVI, 584 pages. 2005.

Vol. 3519: H. Li, P. J. Olver, G. Sommer (Eds.), Computer Algebra and Geometric Algebra with Applications. IX, 449 pages. 2005.

Vol. 3518: T.B. Ho, D. Cheung, H. Liu (Eds.), Advances in Knowledge Discovery and Data Mining. XXI, 864 pages. 2005. (Subseries LNAI).

Vol. 3517: H.S. Baird, D.P. Lopresti (Eds.), Human Interactive Proofs. IX, 143 pages. 2005.

Vol. 3516: V.S. Sunderam, G.D.v. Albada, P.M.A. Sloot, J.J. Dongarra (Eds.), Computational Science – ICCS 2005, Part III. LXIII, 1143 pages. 2005.

Vol. 3515: V.S. Sunderam, G.D.v. Albada, P.M.A. Sloot, J.J. Dongarra (Eds.), Computational Science – ICCS 2005, Part II. LXIII, 1101 pages. 2005.

Vol. 3514: V.S. Sunderam, G.D.v. Albada, P.M.A. Sloot, J.J. Dongarra (Eds.), Computational Science – ICCS 2005, Part I. LXIII, 1089 pages. 2005.

Vol. 3513: A. Montoyo, R. Muñoz, E. Métais (Eds.), Natural Language Processing and Information Systems. XII, 408 pages. 2005.

Vol. 3512: J. Cabestany, A. Prieto, F. Sandoval (Eds.), Computational Intelligence and Bioinspired Systems. XXV, 1260 pages. 2005.

Vol. 3511: U.K. Wiil (Ed.), Metainformatics. VIII, 221 pages. 2005.

Vol. 3510: T. Braun, G. Carle, Y. Koucheryavy, V. Tsaoussidis (Eds.), Wired/Wireless Internet Communications. XIV, 366 pages. 2005.

Vol. 3509: M. Jünger, V. Kaibel (Eds.), Integer Programming and Combinatorial Optimization. XI, 484 pages. 2005.

Vol. 3508: P. Bresciani, P. Giorgini, B. Henderson-Sellers, G. Low, M. Winikoff (Eds.), Agent-Oriented Information Systems II. X, 227 pages. 2005. (Subseries LNAI).

Vol. 3507: F. Crestani, I. Ruthven (Eds.), Information Context: Nature, Impact, and Role. XIII, 253 pages. 2005.

Vol. 3506: C. Park, S. Chee (Eds.), Information Security and Cryptology – ICISC 2004. XIV, 490 pages. 2005.

Vol. 3505: V. Gorodetsky, J. Liu, V. A. Skormin (Eds.), Autonomous Intelligent Systems: Agents and Data Mining. XIII, 303 pages. 2005. (Subseries LNAI).

Vol. 3504: A.F. Frangi, P.I. Radeva, A. Santos, M. Hernandez (Eds.), Functional Imaging and Modeling of the Heart. XV, 489 pages. 2005.

Vol. 3503: S.E. Nikoletseas (Ed.), Experimental and Efficient Algorithms. XV, 624 pages. 2005.

Vol. 3502: F. Khendek, R. Dssouli (Eds.), Testing of Communicating Systems. X, 381 pages. 2005.

Vol. 3501: B. Kégl, G. Lapalme (Eds.), Advances in Artificial Intelligence. XV, 458 pages. 2005. (Subseries LNAI).

Vol. 3500: S. Miyano, J. Mesirov, S. Kasif, S. Istrail, P. Pevzner, M. Waterman (Eds.), Research in Computational Molecular Biology. XVII, 632 pages. 2005. (Subseries LNBI).

Vol. 3499: A. Pelc, M. Raynal (Eds.), Structural Information and Communication Complexity. X, 323 pages. 2005.

Vol. 3498: J. Wang, X. Liao, Z. Yi (Eds.), Advances in Neural Networks – ISNN 2005, Part III. XLIX, 1077 pages. 2005.

Vol. 3497: J. Wang, X. Liao, Z. Yi (Eds.), Advances in Neural Networks – ISNN 2005, Part II. XLIX, 947 pages. 2005.

Vol. 3496: J. Wang, X. Liao, Z. Yi (Eds.), Advances in Neural Networks – ISNN 2005, Part II. L, 1055 pages. 2005.

Vol. 3495: P. Kantor, G. Muresan, F. Roberts, D.D. Zeng, F.-Y. Wang, H. Chen, R.C. Merkle (Eds.), Intelligence and Security Informatics. XVIII, 674 pages. 2005.

Vol. 3494: R. Cramer (Ed.), Advances in Cryptology – EUROCRYPT 2005. XIV, 576 pages. 2005.

Vol. 3493: N. Fuhr, M. Lalmas, S. Malik, Z. Szlávik (Eds.), Advances in XML Information Retrieval. XI, 438 pages. 2005.